Lecture Notes in Computer Science   8357

Commenced Publication in 1973
Founding and Former Series Editors:
Gerhard Goos, Juris Hartmanis, and Jan van Leeuwen

For further volumes:
http://www.springer.com/series/7412

Masakazu Iwamura · Faisal Shafait (Eds.)

Camera-Based Document Analysis and Recognition

5th International Workshop, CBDAR 2013
Washington, DC, USA, August 23, 2013
Revised Selected Papers

 Springer

Editors

Masakazu Iwamura
Graduate School of Engineering
Osaka Prefecture University
Osaka
Japan

Faisal Shafait
The University of Western Australia
Crawley, WA
Australia

ISSN 0302-9743   ISSN 1611-3349 (electronic)
ISBN 978-3-319-05166-6   ISBN 978-3-319-05167-3   (eBook)
DOI 10.1007/978-3-319-05167-3
Springer Cham Heidelberg New York Dordrecht London

Library of Congress Control Number: 2014933578

CR Subject Classification (1998): I.4, I.5, I.7, I.2.7, H.3, H.5, H.4

LNCS Sublibrary: SL6 – Image Processing, Computer Vision, Pattern Recognition and Graphics

Printed on acid-free paper

Springer is part of Springer Science+Business Media (www.springer.com)

Preface

The pervasiveness and wide-spread availability of camera phones and hand-held digital still/video cameras has led the community to recognize document analysis and recognition of digital camera images as a promising and growing sub-field of Document Analysis and Recognition. Constraints imposed by the memory, processing speed, and image quality are leading to new interesting open problems that cannot be directly resolved by traditional techniques.

To cater for the demands of camera-based document processing, the idea of a new satellite workshop of International Conference on Document Analysis and Recognition (ICDAR) was conceived by Prof. Koichi Kise. Together with Prof. David Doermann, he took the responsibility of organizing the first workshop on Camera-Based Document Analysis and Recognition as a satellite workshop of ICDAR 2005 in Seoul, South Korea. The workshop was very well received by the community and hence it was held in 2007 (Curitiba, Brazil), 2009 (Barcelona, Spain), and 2011 (Beijing, China) with the corresponding ICDAR conferences. It is our pleasure to hold the Fifth International Workshop on Camera-Based Document Analysis and Recognition (CBDAR 2013) in Washington D.C., USA, following the success of the past four workshops. The workshop is aimed to provide an opportunity to researchers and developers from various backgrounds to exchange their ideas and explore new research directions through the presentation of recent research activities and discussions.

In the eight years since the first CBDAR was held, the situation surrounding the CBDAR field has been evolving. New technologies have brought a shift in the paradigm from static camera-captured scene image reading to real-time video-based OCR using cameras on wearable devices, possibly complementing the camera input with other sensors (e.g., eye tracking). Such sensors and recent technologies have the potential to understand a user's behavior, habit, and thought, as well as improve user experience while reading.

The program of CBDAR 2013 was organized in a single-track one-day workshop. It consisted of two oral sessions and one poster session. In addition to that, a keynote talk was given by Dr. Kai Kunze from Osaka Prefecture University. Finally, a panel discussion on the state of the art and new challenges was organized as the concluding session of CBDAR 2013.

After the workshop, authors of selected contributions were invited to submit expanded versions of their papers for this edited volume. The authors were encouraged to include the ideas and suggestions that arose during the discussions at the workshop. Thus, this volume contains refereed and improved versions of papers presented at CBDAR 2013. We intend to give a snapshot of state-of-the-art research in the field of camera-based document analysis and recognition.

Finally, we would like to sincerely thank those who are helping to ensure this workshop is a success: Dr. David Doermann (ICDAR General Chair), Prof. Daniel

Lopresti (ICDAR Executive Co-chair), Prof. Apostolos Antonacopoulos (ICDAR Workshop Chair), and other ICDAR organizers for their generous support; the members of the program committee and additional reviewers for reviewing and commenting on all of the submitted papers; IAPR for its sponsorship of the workshop.

The Sixth International Workshop on Camera-Based Document Analysis and Recognition (CBDAR 2015) is planned to be held in Tunis, Tunisia.

December 2013   Masakazu Iwamura
                                                                              Faisal Shafait

Contents

Text Detection and Recognition
in Scene Images

Spatially Prioritized and Persistent Text Detection and Decoding

Hsueh-Cheng Wang[(✉)], Yafim Landa, Maurice Fallon, and Seth Teller

Computer Science and Artificial Intelligence Laboratory,
Massachusetts Institute of Technology, Cambridge, MA 02139, USA
{hchengwang,landa,mfallon,teller}@csail.mit.edu

Abstract. We show how to exploit temporal and spatial coherence to achieve efficient and effective text detection and decoding for a sensor suite moving through an environment in which text occurs at a variety of locations, scales and orientations with respect to the observer. Our method uses simultaneous localization and mapping (SLAM) to extract planar "tiles" representing scene surfaces. Multiple observations of each tile, captured from different observer poses, are aligned using homography transformations. Text is detected using Discrete Cosine Transform (DCT) and Maximally Stable Extremal Regions (MSER), and decoded by an Optical Character Recognition (OCR) engine. The decoded characters are then clustered into character blocks to obtain an MLE word configuration. This paper's contributions include: (1) spatiotemporal fusion of tile observations via SLAM, prior to inspection, thereby improving the quality of the input data; and (2) combination of multiple noisy text observations into a single higher-confidence estimate of environmental text.

Keywords: SLAM · Text detection · Video OCR · Multiple frame integration · DCT · MSER · Lexicon · Language model

1 Introduction

Information about environmental text is useful in many task domains. Examples of outdoor text include house numbers and traffic and informational signage; indoor text appears on building directories, aisle guidance signs, office numbers, and nameplates. Given sensor observations of the surroundings we wish to efficiently and effectively detect and decode text for use by mobile robots or by people (e.g., the blind or visually impaired). A key design goal is to develop text extraction method which is fast enough to support real-time decision-making, e.g., navigation plans for robots and generation of navigation cues for people.

1.1 End-to-End Text Spotting in Natural Scenes

Aspects of end-to-end word spotting have been explored previously. Batch methods for Optical Character Recognition (OCR) have long existed. In a real-time

M. Iwamura and F. Shafait (Eds.): CBDAR 2013, LNCS 8357, pp. 3–17, 2014.
DOI: 10.1007/978-3-319-05167-3_1, © Springer International Publishing Switzerland 2014

Fig. 1. Our approach incorporates Simultaneous Localization and Mapping (SLAM) to combine multiple noisy text observations for further analysis. Top left: three cropped tile observations with decoded characters. Bottom left: the spatial distribution of decoded characters from all observations; each colored dot is a decoded case-insensitive character (legend at right). A clustering step is first used to group decoded characters; each group is shown as a circle around the centroid of the decoded characters. A second clustering step merges characters (circles) into word candidates (rectangles). Next, an optimal word configuration (indicated with line segments) is obtained using a language model. The final outputs are "TABLE" and "ROBOTS" (from source text "Printable Robots").

setting, however, resource constraints dictate that text decoding should occur only in regions that are likely to contain text. Thus, efficient text detection methods are needed. Chen and Yuille [1] trained a strong classifier using AdaBoost to identify text regions, and used commercial OCR software for text decoding.

Neumann and Matas [2–4] used Maximally Stable Extremal Region (MSER) [5] detection and trained a classifier to separate characters from non-characters using several shape-based features, including aspect ratio, compactness, and convex hull ratio. They reported an average run time of 0.3 s on an 800×600 image, achieving recall of 64.7 % in the ICDAR 2011 dataset [6] and 32.9 % in the SVT dataset [7].

Wang and colleagues [7,8] described a character detector using Histograms of Oriented Gradient (HOG) features or Random Ferns, which given a word lexicon can obtain an optimal word configuration. They reported computation times of 15 s on average to process an 800×1200 image. Their lexicon driven

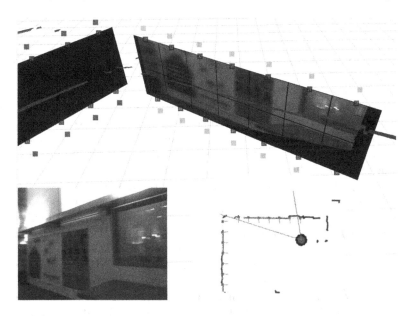

Fig. 2. Top: Visualization of a 3D environment. Each yellow or black label represents an 1×1 m tile. The yellow ones are in camera field of view, and the black ones are discovered by LIDAR, but not by camera. Bottom left: A camera frame. Bottom right: Map generated by the SLAM module (black lines) with generated tiles overlaid (origins in red; normals in green).

method — combining the ABBYY FineReader OCR engine and a state-of-the-art text detection algorithm (Stroke Width Transform (SWT) [9]) — outperformed the method using ABBYY alone.

The open-source OCR engine Tesseract [10,11] has some appealing features, such as line finding, baseline fitting, joined character chopping, and broken character association. Although its accuracy was not as high as that of some other commercial OCR engines [7], it has been widely used in many studies.

1.2   Challenges

We address the problem of extracting useful environmental text from the datastream produced by a body-worn sensor suite. We wish to extract text quickly enough to support real-time uses such as navigation (e.g., the user seeks a numbered room in an office or hotel), shopping (e.g., the user seeks a particular aisle or product), or gallery visits (e.g., the user wants notification and decoding of labels positioned on the walls and floors, or overhead).

To achieve real-time notifications given current network infrastructure, the processing should be performed on-board (i.e., by hardware local to the user), rather than in the cloud, and in a way that exploits spatiotemporal coherence (i.e., the similarity of data available now to data available in the recent past).

First, the user often needs a response in real time, ruling out the use of intermittent or high-latency network connections. Second, the task involves large amounts of data arising from observations of the user's entire field of view at a resolution sufficient for text detection. This rules out reliance on a relatively low-bandwidth network connection. Moreover, in 2013 one cannot analyze a full field of view of high-resolution pixels in real-time using hardware that would be reasonable to carry on one's body (say, a quad- or eight-core laptop). We investigated what useful version of the problem could be solved with wearable hardware, and designed the system to inspect, and extract text from, only those portions of the surroundings that are *newly visible*.

Existing work has incorporated scene text in robotics [12] and assistive technologies for visually impaired or blind people [13]. Unlike scene text in images observed by a stationary camera, text observed by a moving camera will generally be subject to motion blur or limited depth of field (i.e., lack of focus). Blurry and/or low-contrast images make it challenging to detect and decode text. Neither increasing sensor resolution, nor increasing CPU bandwidth, are likely to enable text detection alone; instead, improved methods are required.

For blurry or degraded images in video frames, multi-frame integration has been applied for *stationary* text [14–16], e.g., captions in digital news, and implemented for text enhancement at pixel or sub-pixel level (see [17]). However, additional registration and tracking are required for text in 3D scenes in video imagery [18].

2   The Proposed Method

SLAM has long been a core focus of the robotics community. On-board sensors such as cameras or laser range scanners (LIDARs) enable accurate egomotion estimation with respect to a map of the surroundings, derived on-line. Large-scale, accurate LIDAR-based SLAM maps can now be generated in real time for a substantial class of indoor environments. Incremental scan-matching and sensor fusion methods have been proposed by a number of researchers [19–21]. We incorporate SLAM-based extraction of $1\,m \times 1\,m$ "tiles" to improve text-spotting performance.

Our system uses SLAM to discover newly visible vertical tiles (Fig. 2), along with distance to and obliquity of each scene surface with respect to the sensor. For example, text can be decoded more accurately when the normal of the surface on which it occurs is roughly perpendicular to the viewing direction. Furthermore, a SLAM-based approach can track the reoccurrence of a particular text fragment in successive image frames. Multiple observations can be combined to improve accuracy, e.g., through the use either of super-resolution methods [22,23] to reduce blur before OCR, or probabilistic lexical methods [8,24] to combine the noisy low-level text fragments produced by OCR. The present study focuses on the latter method.

Some designers of text detection methods have used the texture-based Discrete Cosine Transform (DCT) to detect text in video [25,26]. Others have used

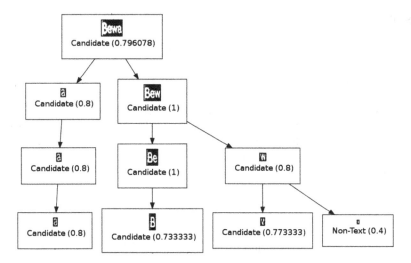

Fig. 3. MSER component tree. Each node was classified as (potential) text, or as nontext, based on shape descriptors including compactness, eccentricity, and the number of outer boundary inflexion points.

MSER, which is fast and robust to blur, low contrast, and variation in illumination, color and texture [4]. We use an 8×8-pixel window DCT as a first-stage scan, then filter by size and aspect ratio. For blurry inputs, individual characters of a word usually merge into one connected component, which could be explored in the component tree generated by MSER [3,27,28]; see Fig. 3. We use MSER with shape descriptors for second-stage classification, to extract individual characters and to produce multiple detection regions for each character, which are then provided to Tesseract.

The availability of multiple observations of each tile enable our method to integrate information (Fig. 1). A clustering process incorporates spatial separation and lexical distance to group decoded characters across multiple frames . Candidate interpretations are combined within each group (representing a single character) using statistical voting with confidence scores. A second clustering step merges groups to form word candidates using a distance function described below.

Extracting environment text from word candidates is similar to the problem of handwriting word recognition, which involves (i) finding an optimal word configuration (segmentation) and (ii) finding an optimal text string. Our approach differs from that of Wang et al. [7,8], who considered (i) and (ii) as a single problem of optimal word configuration using pictorial structure; we separate (i) and (ii) in order to reduce running time and increase control over the individual aggregation stages.

3   System

See Fig. 4 for an overview of our system's dataflow.

3.1   Sensor Data Inputs

Data was collected from a wearable rig containing a Hokuyo UTM-30LX planar LIDAR, a Point Grey Bumblebee2 camera, and a Microstrain 3DM-GX3-25 IMU, shown in Fig. 5. The IMU provides pitch and roll information. All sensor data was captured using the LCM (Lightweight Communications and Marshaling) [29] package.

3.2   Extraction

As the sensor suite moves through the environment, the system maintains an estimate of the sensor rig's motion using incremental LIDAR scan-matching [20] and builds a local map consisting of a collection of line segments (Fig. 2). Two line segments are merged if the difference of their slopes is within a given threshold and offset. Each line segment is split into several 1-m lateral extents which we call tiles. Newly visible tiles are added using the probabilistic Hough transform [30]. For each new tile the system creates four tile corners, each half a meter vertically and horizontally away from the tile center.

3.3   Image Warping

Any tiles generated within the field of view are then projected onto the frames of the cameras that observed them. Multiple observations can be gathered from

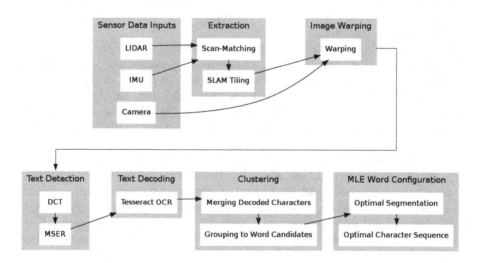

Fig. 4. System dataflow. Our system takes images and laser range data as inputs, extracts tiles, invokes text detection on tiles, and finally schedules text decoding for those tiles on which text was detected.

Fig. 5. The sensors were mounted on a rig and connected to a laptop computer for data collection.

various viewing positions and orientations. A fronto-parallel view of each tile is obtained for each observation through a homography transform constructed by generating a quadrilateral in OpenGL, and using projective texture mapping from the scene image onto the tile quadrilateral. A virtual camera is then placed in front of each tile to produce the desired fronto-parallel view of that tile at any desired resolution (we use 800×800 pixels). The per-tile transform is maintained, enabling alignment of multiple observations in order to later improve image quality and OCR accuracy.

Each individual observation is associated with a tile (its unique identifier, corners, origin, and normal vector), the synthesized fronto-parallel image, and the camera pose. These observations are then passed to text detection and decoding.

3.4 Text Detection

The first stage of text detection applies an image pyramid to each tile in preparation for multi-scale DCT, with coefficients as per Crandall et al. [25]. The bounding box of each text detection is then inspected using MSER [4] to extract shape descriptors, including aspect ratio and compactness. We set the MSER parameters as follows: aspect ratio less than 8, and compactness greater than 15. Scale-relevant parameters are estimated according to real-world setting (8 pixels per cm), corresponding to a minimum text height of 3 cm, and a minimum MSER region of 3 cm^2. The parameters for DCT detection include a minimum edge density of 8 edge-pixels per 8×8 window using Canny edge detection, with high and low hysteresis parameters equal to 100 and 200, respectively. For MSER detection, regions smaller than 5 pixels are discarded, and the parameter delta (the step size between intensity threshold levels) is set to 3 for better robustness to blurry inputs. Both the DCT and MSER computations are implemented in OpenCV, with running times of about 10 ms and 300 ms, respectively.

3.5   Text Decoding

Decoding proceeds as follows. First, the image regions produced by either DCT or MSER (as gray-scale or binary images) are processed by the Tesseract OCR engine. Using the provided joined character chopping and broken character association, the binary inputs are segmented into one or multiple observations, i.e., the segmentation results from a MSER region. Tesseract outputs with too large an aspect ratio are removed. Each block is classified into a few candidates with confidence scores, for example, "B", "E" and "8" for the crop of an image of character "B." We set a minimum confidence score of 65 given by Tesseract to reduce incorrectly decoded characters. Running time depends on the number of input regions, but is usually less than 300 ms.

3.6   Clustering for Character and Word Candidates

A clustering module is used to: (a) merge decoded characters across multiple observations, and (b) cluster groups of decoded characters into word candidates. For (a), our distance function incorporates Euclidean distance, text height, and similarity between decoded results. Multiple observations can be obtained either across multiple frames or within a single frame. The parameters of multi-frame integration depend on system calibration. For (b), the confidence of groups of decoded characters, size of decoded characters, and Euclidean distance are applied. The confidence is determined by the number of decoded characters in the group; only groups with confidence above a threshold are selected. The threshold is $\sqrt{N_{obs}}/k$, where N_{obs} is the total number of accumulated decoded characters with a parameter k set to 1.3 in our application. The bounding box of each decoded character in selected groups are overlaid on a density map, which is then segmented into regions. All selected groups of decoded characters are assigned to a region, representing a word candidate.

3.7   Finding Optimal Word Configuration and String

To extract whole words, we implemented a graph to combine spatial information (block overlaps). The output is a sequence of characters with each character comprising a small number of candidates provided by Tesseract. To recover the optimal word string, each candidate from each group of decoded characters is considered as a node in a trellis, where the probability of each node arises from normalized voting using confidence scores. The prior probability is computed using bi-grams from an existing corpus [31]. We retain the top three candidates for each group of decoded characters, and use Viterbi's algorithm [32] for decoding. We seek an optimal character sequence W^*, as shown in Eq. 1, where $P(Z|C_i)$ is the probability of nodes from the confidence-scored observations, and $P(C_i|C_{i-1})$ is the prior probability from the bi-gram:

$$W^* = \underset{w}{\operatorname{argmax}} \left(\sum P(Z|C_i)P(C_i|C_{i-1}) \right) \qquad (1)$$

4   Experimental Results

Text examples in public datasets (e.g., ICDAR and SVT) usually occur within high-quality (high-resolution, well-focused) imagery. In our setting, text often occurs at lower-resolution and with significant blur. Our focus is to achieve text-spotting in a real-time system moving through an environment. We first examine how much the information about the surround given by SLAM and the warping process affect text detection and decoding in video frames. Next, we demonstrate the alignment of warped tile observations. Finally, we evaluate the accuracy gains arising from spatiotemporal fusion.

The evaluation is performed using a metric defined over m ground truth words and n decoded words. The $m \times n$ pairs of strings are compared using minimum edit distance d_{ij} for the ith ground truth word and the jth decoded word. A score S_{ij} for each pair is calculated as $(N_i - d_{ij})/N_i$, where N_i is the number of character of ground truth word i, when $N_i - d_{ij} > 0$, with S_{ij} set to 0 otherwise. The accuracy is then measured by Eq. 2, where the weight of each ground truth word w_i is set to $1/max(m,n)$ to penalize false alarms when $n > m$.

$$\text{Accuracy} = \sum_i w_i \max_j (S_{ij}) \qquad (2)$$

4.1   Warping Accuracy with Distance and Obliquity

We mounted all equipment on a rig placed at waist height on a rolling cart, with the LIDAR sampling at 40 Hz and the camera sampling at 15 Hz. We attached signs with text in a 140-point (5 cm) font at various wall locations. We pushed the cart slowly toward and by each sign to achieve varying view angles with respect to the sign's surface normal (Fig. 6(a, b)). The experiments were designed to evaluate text-spotting performance under varying viewing distance and obliquity, given that such factors effect the degree of blur.

Each original tile and its warped observation cropped from the image frame was sent to Tesseract, our baseline decoder. Text spotting performance vs. the baseline is plotted as a function of viewing distance (Fig. 6(c, d)). Examples are shown in Fig. 6(e, f).

The results suggest that the baseline decoder works poorly when text is observed at distances greater than 1.5 m, and generally performs better for the warped observation than for the original ones. When the viewing direction is about 45° from the surface normal, the text extracted from the warped images is more consistent than that from the unwarped images, which may be due to the skewed text line and perspective transformation of characters in the latter.

Given the limitations of the baseline decoder, our proposed method extends the capability of detecting and decoding lower-quality imagery through spatiotemporal fusion of multiple observations. One key factor for integration is: how well are the warped observations aligned? We describe our use of alignment in the next section.

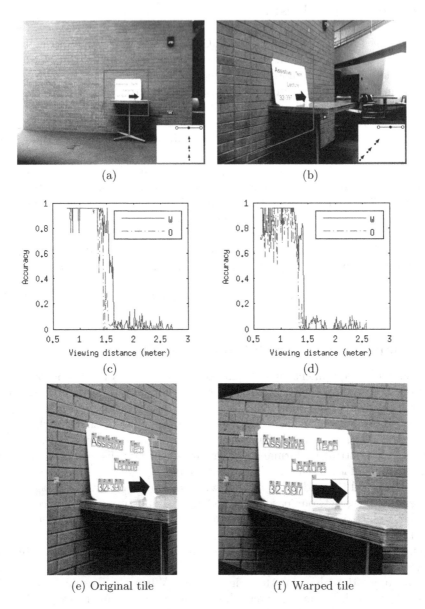

(a)   (b)

(c)   (d)

(e) Original tile   (f) Warped tile

Fig. 6. Experimental settings and accuracy comparison of original and warped observations. (a) The normal of the surface is roughly antiparallel to the viewing direction. (b) The normal of the surface is about 45° away from the viewing direction. Plots (c) and (d) show the accuracy of baseline decoding of original (O) and warped (W) tiles with respect to viewing distance for observations (a) and (b). (e) An original tile observation from 0.71 m. (f) The warped observation corresponding to (e). The accuracy scores of (e) and (f) are 0.67 and 0.96, respectively.

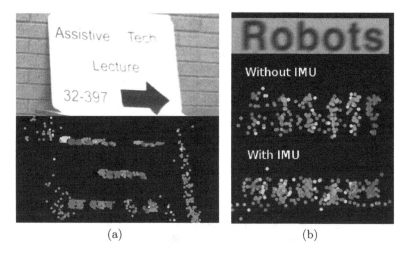

(a)   (b)

Fig. 7. The distribution of decoded characters. (a) There were only slight vertical and horizontal shifts. (b) Comparison between data with and without IMU for the second dataset (hand-carried). There were longer vertical drifts without IMU, but use of the IMU reduces drift.

4.2   Alignment of Warped Observations

The distribution of decoded characters is shown in Fig. 7(a, b). Misalignment among certain decoded characters was measured manually. In Fig. 7(a), the logs were collected when the sensors were placed on a cart. The results suggest that the drift of decoded characters was uncertain to within about 20 pixels.

Another log was collected when the rig was hand-carried at about chest height by an observer who walked within an indoor environment. Figure 7(b) demonstrates that imagery, to be aligned, required shifts of around 20 pixels horizontally and 70 pixels vertically without IMU data. When IMU data were integrated, the vertical shifts required reduced to around 35 pixels.

Given the alignment, we used the configuration of Fig. 6(b) to study the text-spotting performance of fusion of multiple observations. The parameter settings for clustering decoded characters and word candidates are shown in Table 1. Comparing single and multiple frame integrations, Euclidean distance is the major factor for merging decoded characters, whereas the threshold of the number of decoded character per group is the major factor for grouping to word candidates.

4.3   Performance with Multiple Observations

We demonstrate that the proposed method combines noisy individual observations into a higher-confidence decoding. Figure 8 plots the accuracy of (a) running Tesseract on the entire tile observation (Tess), (b) combining (a) and the proposed spotting pipeline into a single-frame detector (Tess+DMTess), and

Table 1. Parameters for character and word clustering.

	Single frame	Multiple frames
Merge decoded characters		
Euclidean distance	10	30
Text height scalar	2	2
Decoded text similarity	1	1
Group to word candidates		
Threshold of characters per group	1	$\sqrt{N_{obs}}/k$
Threshold parameter k		1.3
Size outlier scalar	5	2
Text height outlier scalar	5	2
Characters per word	3	3
Word aspect ratio min	1	1
Bounding box horizontal increment	0.3	0.3
Bounding box vertical increment	0.05	0.05

(c) fusing multiple observations from the proposed pipeline (Multi). The area under curve (AUC) values are 0.71, 0.79, and 0.91, respectively; these represent the overall performance of each spotting pipeline. The results suggest that Tess+DMTess moderately extends (from 1.5 to 2.4 m) the distance at which text can be decoded, and Multi moderately improves the accuracy with which blurry text can be decoded (since blur tends to increase with viewing distance). We found that reducing the rate of false positives is critical to successful fusion, because a high false-alarm rate tends to cause our clustering method (Sect. 3.6) to fail. We will continue to investigate our observation that Tess+DMTess outperforms Multi for close observations (1–1.5 m).

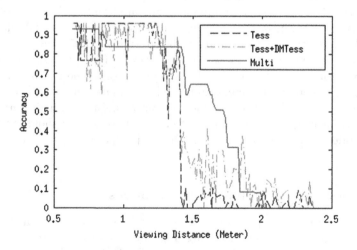

Fig. 8. Accuracy with respect to viewing distance for observations.

5   Conclusion and Future Work

We described a SLAM-based text spotting method which detects and decodes scene text by isolating "tiles" arising from scene surfaces observed by a moving sensor suite. Such mode of operation poses challenges to conventional text detection methods using still imagery or stationary video frames. We demonstrate how SLAM-derived information about the surroundings can be used to improve text spotting performance. We also show how to merge text extracted from multiple tile observations, yielding higher-confidence word recovery end-to-end. Our future work will (1) incorporate a more sophisticated tile orientation and camera motion model into the observation alignment, clustering, and language model; (2) collect large-scale datasets for evaluation; and (3) schedule computationally intensive inspection according to a spatial prior on text occurrence to improve efficiency. Finally, we plan to explore the use of the method to support task performance in robotics and assistive technology for blind and visually impaired people.

Acknowledgment. We thank the Andrea Bocelli Foundation for their support, and Javier Velez and Ben Mattinson for their contributions.

References

1. Chen, X., Yuille, A.: Detecting and reading text in natural scenes. In: Proceedings of the IEEE International Conference on Computer Vision and Pattern Recognition (CVPR) (2004)
2. Neumann, L., Matas, J.: A method for text localization and recognition in real-world images. In: Asian Conference on Computer Vision (ACCV), pp. 770–783 (2004)
3. Neumann, L., Matas, J.: Text localization in real-world images using efficiently pruned exhaustive search. In: Proceedings of the International Conference on Document Analysis and Recognition (ICDAR), pp. 687–691 (2011)
4. Neumann, L., Matas, J.: Real-time scene text localization and recognition. In: Proceedings of the IEEE International Conference Computer Vision and Pattern Recognition (CVPR) (2012)
5. Matas, J., Chum, O., Urban, M., Pajdla, T.: Robust wide-baseline stereo from maximally stable extremal regions. Image Vis. Comput. **22**(10), 761–767 (2004)
6. Lucas, S.: ICDAR 2005 text locating competition results. In: Proceedings of the International Conference on Document Analysis and Recognition (ICDAR), vol. 1, pp. 80–84 (2005)
7. Wang, K., Belongie, S.: Word spotting in the wild. In: Daniilidis, K., Maragos, P., Paragios, N. (eds.) ECCV 2010, Part I. LNCS, vol. 6311, pp. 591–604. Springer, Heidelberg (2010)
8. Wang, K., Babenko, B., Belongie, S.: End-to-end scene text recognition. In: International Conference on Computer Vision (ICCV) (2011)
9. Epshtein, B., Ofek, E., Wexler, Y.: Detecting text in natural scenes with stroke width transform. In: Proceedings of the IEEE International Conference Computer Vision and Pattern Recognition (CVPR), pp. 2963–2970 (2010)

10. Smith, R.: An overview of the tesseract OCR engine. In: Proceedings of the International Conference on Document Analysis and Recognition (ICDAR), pp. 629–633 (2007)
11. Smith, R.: History of the tesseract OCR engine: what worked and what didn't. In: Proceedings of SPIE Document Recognition and Retrieval (2013)
12. Posner, I., Corke, P., Newman, P.: Using text-spotting to query the world. In: IEEE/RSJ International Conference on Intelligent Robots and Systems (IROS), pp. 3181–3186 (2010)
13. Yi, C., Tian, Y.: Assistive text reading from complex background for blind persons. In: Proceedings of Camera-based Document Analysis and Recognition (CBDAR), pp. 15–28 (2011)
14. Sato, T., Kanade, T., Hughes, E., Smith, M.: Video OCR for digital news archive. In: Proceedings 1998 IEEE International Workshop on Content-Based Access of Image and Video Database, pp. 52–60 (1998)
15. Li, H., Doermann, D.: Text enhancement in digital video using multiple frame integration. In: Proceedings of the seventh ACM international conference on Multimedia (Part 1), pp. 19–22 (1999)
16. Hua, X.S., Yin, P., Zhang, H.J.: Efficient video text recognition using multiple frame integration. In: Proceedings of the 2002 International Conference on Image Processing, vol. 2 II-397–II-400 (2002)
17. Jung, K., Kim, K.I., Jain, A.K.: Text information extraction in images and video: a survey. Pattern Recogn. **37**(5), 977–997 (2004)
18. Myers, G.K., Burns, B.: A robust method for tracking scene text in video imagery. In: CBDAR05 (2005)
19. Olson, E.: Real-time correlative scan matching. In: IEEE International Conference on Robotics and Automation (ICRA), Kobe, Japan, pp. 4387–4393, June 2009
20. Bachrach, A., Prentice, S., He, R., Roy, N.: RANGE - robust autonomous navigation in GPS-denied environments. J. Field Robot. **28**(5), 644–666 (2011)
21. Fallon, M.F., Johannsson, H., Brookshire, J., Teller, S., Leonard, J.J.: Sensor fusion for flexible human-portable building-scale mapping. In: IEEE/RSJ International Conference on Intelligent Robots and Systems (IROS), Algarve, Portugal (2012)
22. Park, S.C., Park, M.K., Kang, M.G.: Super-resolution image reconstruction: a technical overview. IEEE Signal Process. Mag. **20**(3), 21–36 (2003)
23. Farsiu, S., Robinson, M., Elad, M., Milanfar, P.: Fast and robust multiframe super resolution. IEEE Trans. Image Process. **13**(10), 1327–1344 (2004)
24. Mishra, A., Alahari, K., Jawahar, C.: Top-down and bottom-up cues for scene text recognition. In: Proceedings of the IEEE International Conference Computer Vision and Pattern Recognition (CVPR), pp. 2687–2694 (2012)
25. Crandall, D., Antani, S., Kasturi, R.: Extraction of special effects caption text events from digital video. Int. J. Doc. Anal. Recogn. **5**(2–3), 138–157 (2003)
26. Goto, H.: Redefining the dct-based feature for scene text detection. Int. J. Doc. Anal. Recogn. (IJDAR) **11**(1), 1–8 (2008)
27. Nistér, D., Stewénius, H.: Linear time maximally stable extremal regions. In: Forsyth, D., Torr, P., Zisserman, A. (eds.) ECCV 2008, Part II. LNCS, vol. 5303, pp. 183–196. Springer, Heidelberg (2008)
28. Merino-Gracia, C., Lenc, K., Mirmehdi, M.: A head-mounted device for recognizing text in natural scenes. In: Proceedings of Camera-based Document Analysis and Recognition (CBDAR), pp. 29–41 (2011)
29. Huang, A., Olson, E., Moore, D.: LCM: Lightweight communications and marshalling. In: IEEE/RSJ International Conference on Intelligent Robots and Systems (IROS), Taipei, Taiwan, October 2010

30. Bonci, A., Leo, T., Longhi, S.: A Bayesian approach to the Hough transform for line detection. IEEE Trans. Syst. Man Cybern. Part A: Syst. Hum. **35**(6), 945–955 (2005)
31. Jones, M.N., Mewhort, D.J.K.: Case-sensitive letter and bigram frequency counts from large-scale english corpora. Behav. Res. Meth. Instrum. Comput. **36**(3), 388–396 (2004)
32. Rabiner, L.: A tutorial on hidden Markov models and selected applications in speech recognition. Proc. IEEE **77**(2), 257–286 (1989)

A Hierarchical Visual Saliency Model for Character Detection in Natural Scenes

Renwu Gao[1]([✉]), Faisal Shafait[2], Seiichi Uchida[3], and Yaokai Feng[3]

[1] Information Sciene and Electrical Engineering, Kyushu University, Fukuoka, Japan
kou@human.ait.kyushu-u.ac.jp
[2] The University of Western Australia, Perth, Australia
[3] Kyushu University, Fukuoka, Japan

Abstract. Visual saliency models have been introduced to the field of character recognition for detecting characters in natural scenes. Researchers believe that characters have different visual properties from their non-character neighbors, which make them salient. With this assumption, characters should response well to computational models of visual saliency. However in some situations, characters belonging to scene text mignt not be as salient as one might expect. For instance, a signboard is usually very salient but the characters on the signboard might not necessarily be so salient globally. In order to analyze this hypothesis in more depth, we first give a view of how much these background regions, such as sign boards, affect the task of saliency-based character detection in natural scenes. Then we propose a hierarchical-saliency method for detecting characters in natural scenes. Experiments on a dataset with over 3,000 images containing scene text show that when using saliency alone for scene text detection, our proposed hierarchical method is able to capture a larger percentage of text pixels as compared to the conventional single-pass algorithm.

Keywords: Scene character detection · Visual saliency models · Saliency map

1 Introduction

Detection of characters in natural scenes is still a challenging task. One of the reasons is the complicated and unpredictable backgrounds. Another reason is the variety of the character fonts. Many methods have been proposed with the hope of solving the above problems. Coates *et al.* [1] employed a large-scale unsupervised feature learning algorithm to solve the blur, distortion and illumination effects of fonts. Yao *et al.* [2] proposed a two-level classification scheme to solve the arbitrary orientation problem. Mishra *et al.* [3] presented a framework, in which the Conditional Random Field model was used as bottom up cues, and a lexicon-based prior was used as top down cues. Li *et al.* [4] employed adaboost algorithm to combine six types of feature sets. Epshtein *et al.* [5] use the Stroke Width Transform (SWT) feature, which is able to detect characters regardless

M. Iwamura and F. Shafait (Eds.): CBDAR 2013, LNCS 8357, pp. 18–29, 2014.
DOI: 10.1007/978-3-319-05167-3_2, © Springer International Publishing Switzerland 2014

Fig. 1. Examples of salient objects (bounded by red lines) containing characters in natural scenes (Color figure online).

of its scale, direction, font and language. In addition to those recent trials, many methods have been proposed [6].

Some other researchers have tried to employ visual attention models as features [7]. In the recent years, visual attention models have been employed for various object detection/recognition tasks [8–10]. Though the usage of visual attention models for character detection is still under-investigated, their effectiveness has been shown by Shahab *et al.* [11,12] and Uchida *et al.* [13]. Those researchers, who try to employ visual attention models for scene character detection, believe that the characters have different properties compared with their non-character neighbors (*pop-out*). This assumption is acceptable considering that the characters in natural scenes, such as those in Fig. 1 are used to convey "important" information *efficiently* to the passengers.

In some situations, characters are not salient when calculated by saliency-based method; instead, regions in which the characters are written are salient. However, when we only focus on those regions, characters become salient.

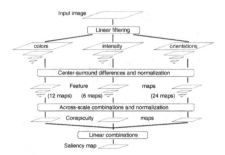

Fig. 2. A general architecture of Itti et. al.'s model. Reference from [14].

In this paper, we investigated how much those regions affect the task of character detection in natural scenes using visual saliency models. We also made a new assumption according to the investigation. The key contribution of this paper is the proposal of a new method for character detection and, compared to the conventional method, the proposed method obtained a better result.

2   Visual Saliency Models

In 1998, Itti *et al.* [14] proposed the first complete implementation and verification of the Koch & Ullman visual saliency model [15]. After that, several kinds of saliency models were proposed [16]. The visual attention models, most of which are directly or indirectly inspired by the mechanism and the neuronal architecture of the primate visual system, are studied to simulate the behavior of human vision [14]. These models provide a massively parallel method for the selection of the intesting objects for the later processing. Visual attention models have been applied to predict where we are focusing in a given scene (an image or a video frame) [17].

2.1   The Calculation of Saliency Map

Many implementations of visual saliency models have been proposed. In this paper, we employ the Itti *et al.*'s model [14] to detect characters in natural scenes. As shown in Fig. 2, three channels (Intensity, Color and Orientation) are used as the low level features [18] to calculate the saliency map as follows:

1. Feature maps are calculated for each channel via center-surround differences operation;
2. Three kinds of conspicuity maps are obtained by across-scale combination;
3. The final saliency map is built through combining all of the conspicuity maps.

Figure 3 shows saliency maps of scene images of Fig. 1, by Itti *et al.*'s models. All the visual saliency maps, in this paper, are calculated using Neuromorphic Vision C++ Toolkit (iNVT), which is developed at iLab, USC [19].

2.2 The Problem of Using Saliency Map for Scene Character Detection

From Fig. 3(a), we can find that the characters are salient as we expected. However, in the cases of Fig. 3(b) – (e), pixels belonging to the characters themselves are less salient as we expected; instead, objects (such as the signboards) containing those characters are salient enough to attract our attention.

The examples of Fig. 3 reveal that, in some situation, characters are not salient if we review the whole image; however, when we focus on the signboards, the characters become conspicuous. This means that signboards (or other objects on which the characters are written) are designed to be salient globally (compared to other parts of the image), whereas the characters are designed to be salient locally (compared to their surrounding region).

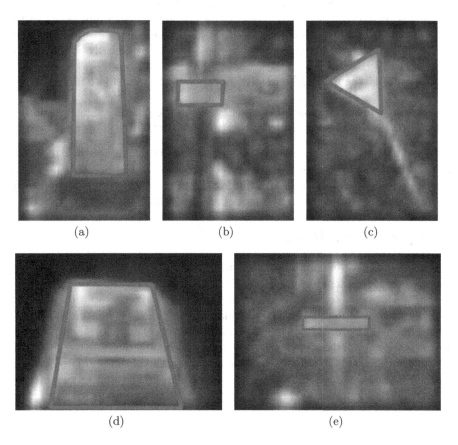

(a)   (b)   (c)

(d)   (e)

Fig. 3. The corresponding saliency maps of Fig. 1. The surrounding regions are bounded by red lines (Color figure online).

2.3 A Hierarchical-Saliency Model

Based on the above observation, we now have a new assumption: characters are prominent compared to their near non-character neighbors, although they may not be so in a global view of the image. In other words, characters are often *locally salient* inside their possibly *globally salient* surrounding region. For example, characters on a car license number plate may be less prominent than the license number plate when we look at the entire car, but they become prominent if we only look at the number plate.

Correspondingly, a new approach for detecting characters in natural scenes is proposed in this study (called the hierarchical-saliency method) which is briefly introduced below:

– First step (extraction of globally salient region):
  1. A saliency map S is calculated from input image I;
  2. The regions of interest (ROIs) of S are evaluated (the procedure of the evaluation will be provided later) and all pixels are automatically classified into two categories to obtain mask M: the globally salient region (1) and the rest (0);
  3. Multiply the mask M with the input image I to calculate filtered image I';
– Second step (evaluation of local saliency inside the globally salient region): Use I' to obtain a new saliency map S', which is the final map we want.

It is very important to note that though we use the same saliency model to calculate the saliency map in both first and second step, the first saliency value and the second value are different even for the same characters. This is simply because the areas subjected to the model are different.

3   Experimental Results

Two experiments were included: (1) in order to investigate how much the salient regions where characters were written affect the task of scene character detection, we firstly arbitrarily selected 101 images from the database and cropped them manually, then calculated the saliency maps for all the 101 images using Itti's saliency model; (2) in order to give a comparison of the performance between the conventional method and the hierarchical-saliency method, we used the whole database with 3,018 images to calculate both the *global* and *local* saliency map, and the salient regions were automatically cropped using Otsu's method and/or Ward's hierarchical clustering method in the process of extracting the ROI regions.

3.1 Database

The scenery image database containing 3,018 images of different sizes has been prepared by our laboratory[1]. All these images were collected from the website

---

[1] We are planning to make the database freely available in near feature.

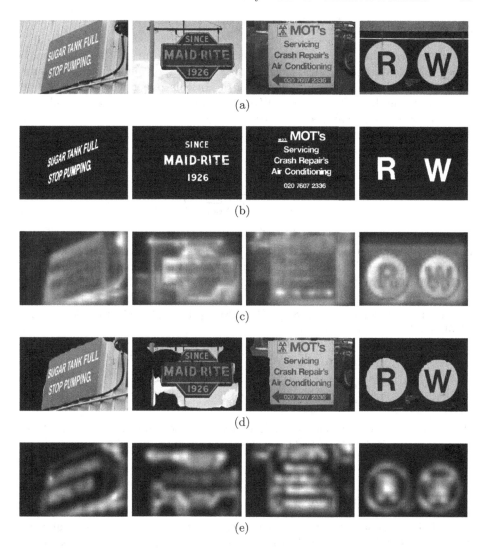

Fig. 4. (a) input images; (b) ground-truth images; (c) Itti *et al.*'s visual saliency maps, calculated using the whole images (Intensity, Color and Orientation); (d) cropped ROI images calculated with (c); (e) Itti *et al.* 's visual saliency maps, calculated within (d) (Intensity, Color and Orientation) (Color figure online).

"flickr". For each image of our database, pixels of characters were labeled in the corresponding image (ground truth image) and the character information (for example, the bounding-box of the character) was stored into a separate text file.

3.2 Extraction of the ROI

How to extract the ROI (signboards, banners, etc.) from the global saliency map S for calculating the local saliency map is an important problem, because the results of the second step depends on it. In this paper, the Otsu's global thresholding method [20] and the Ward's hierarchical clustering method [21] were employed for a trial (see Fig. 4(d)).

In the Ward's hierarchical clustering method, the error sum of squares (ESS) was given as a loss function F:

$$F = \sum_{i=1}^{n} x_i^2 - \frac{1}{n} \left(\sum_{i=1}^{n} x_i \right)^2$$

where x_i is the score of the ith individual and n donates the number of the individulas in the set. This method reduces n sets to $n - 1$ mutually exclusive sets by considering the union of all possible pairs and selecting a union having a minimal value for the loss function F. Assume there are 3 numbers: $\{1,2,8\}$ and we want to group them into 2 sets. In the Ward's method, all the combinations are considered: $\{(1,2),(8)\}$, $\{(1),(2,8)\}$, $\{(1,8),(2)\}$. Then the loss F are calculated for each combination:

$$F_{\{(1,2),(8)\}} = F_{\{(1,2)\}} + F_{\{(8)\}} = 0.5 + 0 = 0.5$$
$$F_{\{(1),(2,8)\}} = F_{\{(1)\}} + F_{\{(2,8)\}} = 0 + 18 = 18$$
$$F_{\{(1,8),(2)\}} = F_{\{(1,8)\}} + F_{\{(2)\}} = 24.5 + 0 = 24.5$$

The combination which made the minimal value of loss function is selected, so the final result is $\{(1,2),(8)\}$. This process is repeated until k groups remain. (please refer to [21] for more details).

3.3 Evaluation Protocol

In the first experiment, we used three low level channels to calculate the saliency map S. While doing the second experiment, in the first step, we also used three low level channels to calculate the saliency map. However, in the second step, saliency map was calculated using different combinations (7 kinds) of channels for each image I, with the purpose of figuring out the best features for character detection. Thresholds t_n ($n \in [0, 255]$) from 0 to 255 were obtained by step 1. Given the corresponding ground truth image I_{GT} with the number of character pixels G_T and the number of non-character pixels G_B, thresholds t_n were applied to evaluate:

1. The number of pixels that matches between saliency map I' (salient pixels) and ground truth I_{GT} (character pixels), $|S_T|$
2. The number of pixels that are salient in the saliency map I', but belong to the non-character regions in the ground-truth image I_{GT}, $|S_B|$.

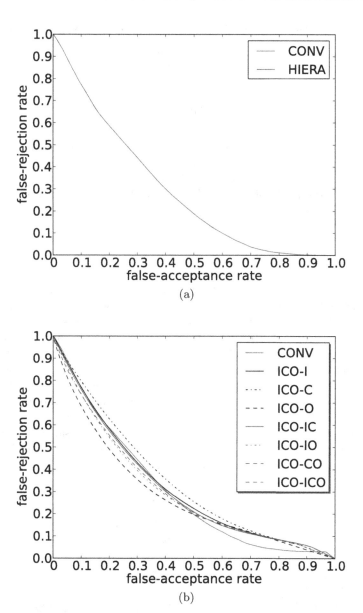

Fig. 5. ROC curves performance comparison. (a) the conventional method (*CONV*) vs. the proposed hierarchical-saliency method with manually cut images; *CONV* represents the conventional method, in which method Itti's model is used only once (first step in our method), *HIERA* represents our proposed method. (b) a comparison between the conventional method and the hierarchical-saliency method with Ward's method cut images (In the second step, we applied all the combination of the low level features); *I / C / O* represent the low level features (*Intensity / Color / Orientation*). Using Otsu's global thresholding method instead of Ward's method gave similar results.

For each threshold, the following performance metrics were calculated:

$$FAR = \frac{|S_B|}{|G_B|} \qquad (1)$$

and

$$FRR = \frac{|G_T| - |S_T|}{|G_T|} \qquad (2)$$

Receiver operator characteristic (ROC) curves were employed to evaluate the performance. Figure 5(a) shows the result of comparison. False acceptance rate (FAR) and false rejection rate (FRR) are plotted on the x and the y axis respectively for the range of threshold values. In Fig. 5(a), the closest curve line to the origin represents the best performance algorithm since it has the lowest equal error rate.

3.4 Results and Discussion

According to Fig. 5(a), we can clearly observe that, compared to the conventional method (using Itti's saliency model once), the proposed hierarchical-saliency method has a better performance. This indicates that the assumption we made in this paper is acceptable. In this section, we give a brief explanation to this. In order to investigate the reason of this result, we built histograms of the true positive pixels for both methods with feature combinations (see Fig. 6). From Fig. 6, we can find that, at the high threshold side, the pixels of characters detected by the hierarchical-saliency method are more salient compared to those detected by the conventional method. On the other hand, the non-salient regions, most of which are non-characters, are suppressed by cropping those salient objects.

From Fig. 5(b) we can see that using orientation as the low level feature in the second step for scene character detection produced the best results. This is

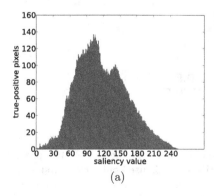
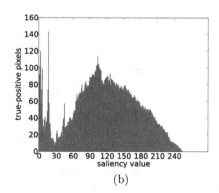

(a)   (b)

Fig. 6. Histograms of the true positive pixels where x-axis represents threshold to decide whether pixels belong to character, and y-axis represents the average number of true positive pixels per one image. (a) histogram calculated using conventional method; (b) histogram calculated using the proposed hierarchical-saliency method

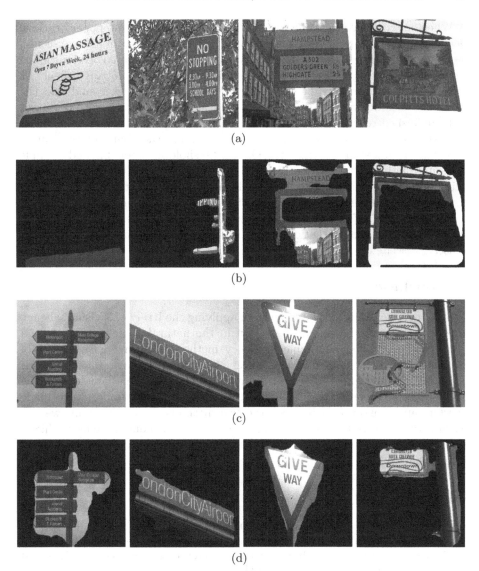

Fig. 7. Examples of the fail and successful images for cropping the surrounding regions. (a) the input images result in the failed results; (b) failed results calculated from (a); (c) the input images result in the successful results; (d) successful results calculated from (c).

mainly because the background in the ROIs is generally simple with few orientation features, whereas the characters have strong orientation features (such as edge features). This makes the characters respond better to the orientation-based visual saliency model and they are easier to detect.

When only using color as the low level feature in the second step, performance became the worst. A possible explanation for this effect is that in natural scenes, the character carriers (the ROIs) are usually designed to be colorful with the purpose of attracting people's attention, which makes them globally salient. As a result, in the procedure of the second step, both the background and the characters respond well to the visual saliency models. Hence characters cannot be distinguished reliably from the background based on saliency alone.

A key issue in our method is how to determine the shape of the salient objects in the first step. We employed Otsu's thresholding algorithm and a simple clustering method (the Ward's hierarchical clustering method) and compared their performance with the conventional method. Though the results of both Otsu' and Ward's method for cropping the salient object were not always good (please refer to Fig. 7 for some successful and failed examples), we still got a better result than the conventional method. It is believable that our method can be used for scene character detection.

4   Conclusion

In this paper, we discussed the problem of applying the Itti *et al.*'s visual saliency model to the task of character detection in the natural scenes, and proposed a new method (called hierarchical-saliency method). We first gave a view of how much the surrounding regions affect character detection, then proposed the Otsu's method and the Ward's hierarchical clustering method to crop the salient objects. In order to investigate the validity of our proposal, we made a performance comparison between the two methods. From the result we can conclude that though the clustering method is not good enough, the hierarchical-saliency method (using orientation feature in the second step) still achieved a better result.

References

1. Coates, A., Carpenter, B., Case, C., Satheesh, S., Suresh, B., Wang, T., Wu, D., Ng, A.: Text detection and character recognition in scene images with unsupervised feature learning. In: International Conference on Document Analysis and Recognition (ICDAR), pp. 440–445 (2011)
2. Yao, C., Bai, X., Liu, W., Tu, Z.: Detection texts of arbitrary orientations in natural images. In: Computer Vision and Pattern Recognition (CVPR), pp. 1083–1090 (2012)
3. Mishra, A., Alahari, K., Jawahar, C.V.: Top-down and bottom-up cues for scene text recognition. In: International Conference on Computer Vision and Pattern Recognition (CVPR), pp. 2687–2694 (2012)
4. Lee, J.J., Lee, P.H., Lee, S.W., Yuille, A., Koch, C.: AdaBoost for text detection in natural scene. In: International Conference on Document Analysis and Recognition (ICDAR), pp. 429–434 (2011)
5. Epshtein, B., Ofek, E., Wexler, Y.: Detecting text in natural scenes with stroke width transform. In: Computer Vision and Pattern Recognition (CVPR), pp. 2963–2970 (2010)

6. Uchida, S.: Text localization and recognition in images and video. In: Doerman, D., Tombre, K.(eds.) Handbook of Document Image Processing and Recognition (to be published in 2013)

7. Sun, Q.Y., Lu, Y., Sun, S.L.: A visual attention based approach to text extraction. In: International Conference on Pattern Recognition (ICPR), pp. 3991–3995 (2010)

8. Walther, D., Itti, L., Riesenhuber, M., Poggio, T.A., Koch, Ch.: Attentional selection for object recognition - a gentle way. In: Bülthoff, H.H., Lee, S.-W., Poggio, T.A., Wallraven, Ch. (eds.) BMCV 2002. LNCS, vol. 2525, pp. 472–479. Springer, Heidelberg (2002)

9. Elazary, L., Itti, L.: A Bayesian model for efficient visual search and recognition. Vision. Res. **50**(14), 1338–1352 (2010)

10. Torralba, A., Murphy, K.P., Freeman, W.T., Rubin, M.A.: Context-based vision system for place and object recognition. In: International Conference on Computer Vision (ICCV), vol. 1, pp. 273–280 (2003)

11. Shahab, A., Shafait, F., Dengel, A.: Bayesian approach to photo time-stamp recognition, In: International Conference on Document Analysis and Recognition (ICDAR), pp. 1039–1043 (2011)

12. Shahab, A., Shafait, F., Dengel, A., Uchida, S.: How salient is scene text?. In: International Workshop on Document Analysis Systems (DAS), pp. 317–321 (2012)

13. Uchida, S., Shigeyoshi, Y., Kunishige, Y., Feng, Y.K.: A keypoint-based approach toward scenery character detection. In: International Conference on Document Analysis and Recognition (ICDAR), pp. 918–823 (2011)

14. Itti, L., Koch, C., Niebur, E.: A model of saliency-based visual attention for rapid scene analysis. IEEE Trans. Pattern Anal. Mach. Intell. (PAMI) **20**(11), 1254–1259 (1998)

15. Koch, C., Ullman, S.: Shifts in selective visual attention: towards the underlying neural circuitry. Human Neurobiol. **4**, 219–227 (1985)

16. Borji, A., Itti, L.: State-of-the-art in visual attention modeling. IEEE Trans. Pattern Anal. Mach. Intell. (PAMI) **35**(1), 185–207 (2013)

17. Judd, T., Ehinger, K., Durand, F., Torralba, A.: Learning to predict where humans look. In: International Conference on Computer Vision, Kyoto, Japan, pp. 2016–2113 (2009)

18. Treisman, A.M., Gelade, G.: A feature-integration theory of attention. Cogn. Psychol. **12**(1), 97–136 (1980)

19. http://ilab.usc.edu/toolkit

20. Otsu, N.: A threshold selection method from gray-level histograms. IEEE Trans. Sys. Man Cybern. **9**(1), 62–66 (1979)

21. Ward Jr, J.H.: Hierarchical grouping to optimize an object function. J. Am. Stat. Assoc. **58**(301), 236–244 (1963)

A Robust Approach to Extraction of Texts from Camera Captured Images

Sudipto Banerjee[1], Koustav Mullick[1], and Ujjwal Bhattacharya[2(✉)]

[1] Department of Computer Science and Engineering, Heritage Institute of
Technology, Kolkata, India
[2] Computer Vision and Pattern Recognition Unit, Indian Statistical Institute,
Kolkata, India
ujjwal@isical.ac.in

Abstract. Here, we present our recent study of a robust but simple
approach to extraction of texts from camera-captured images. In the
proposed approach, we first identify pixels which are highly specular.
Connected components of this set of specular pixels are obtained. Pixels belonging to each such component are separately binarized using the
well-known Otsu's approach. We next apply smoothing on the whole
image before obtaining its Canny edge representation. Bounding rectangle of each connected component of the Canny edge image is obtained
and multiple components with pairwise overlapping bounding boxes are
merged. Otsu's thresholding technique is applied separately on different
parts of input image defined by the resulting bounding boxes. Although
Otsu's thresholding approach does not generally provide acceptable performance on camera captured images, we observed its suitability when
applied severally as in the above. The binarized specular components
obtained at the initial stage replace the corresponding regions of the latter binarized image. Finally, a set of postprocessing operations is used
to remove certain non-text components of the binarized image.

1 Introduction

Recognition of texts in natural scene images has several important applications
such as reading aids for tourists travelling a place with foreign script or for the
blinds. Extraction and binarization of texts in scene images is an important step
for this recognition problem. Traditional binarization approaches belong to two
categories: global thresholding methods [1,2] and local thresholding methods
[3,4]. A recent evaluation study of existing thresholding methods for document
images can be found in [5]. However, these methods often suffer from poor performance on natural scene text images affected by numerous degradations such
as uneven lighting, complex background, low contrast, multiple colours in both
foreground and background etc. Recently, Peng et al. [6] studied an MRF based
binarization approach for hand-held device captured document images which
also cannot be directly applied to such images due to the above challenges in
addition to the problems of enormous variability in fonts of texts.

M. Iwamura and F. Shafait (Eds.): CBDAR 2013, LNCS 8357, pp. 30–46, 2014.
DOI: 10.1007/978-3-319-05167-3_3, © Springer International Publishing Switzerland 2014

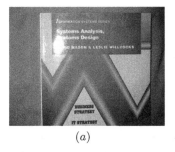
(a)

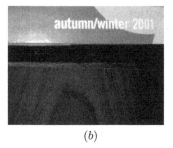
(b)

Fig. 1. (a) A sample image containing a large glare of light falling on the text region, (b) a sample image with varying contrast

A robust but simple method to extract as much texts as possible from natural scene images affected by serious uneven lighting conditions such as specular reflection of light as in the sample image shown in Fig. 1(a) or varying contrasts as shown in Fig. 1(b) is proposed in this paper. Our aim is to enhance and extract the information after reducing the presence of noisy pixels as much as possible. The proposed approach not only extracts texts from scene images having uneven specular reflections, in which most of the current algorithms fail, but also works equally well for other difficult situations such as presence of shadow, varying contrasts, texts at arbitrary orientations etc. Here, we use Otsu's thresholding method [1] for binarization purposes. It is an established fact that Otsu's method is an efficient binarization approach as long as the input image is not affected by non-uniform illumination. Thus, we apply Otsu's method severally on smaller rectangular regions within each of which the variation in illumination condition is less likely to occur. Moreover, the smaller rectangular regions are not selected on an ad hoc basis but these are obtained as the minimum rectangular regions enclosing the connected components of Canny edge image. The above is the main factor behind the robustness of the proposed approach for extraction of texts from general camera captured images.

Rest of this article is organized as follows. Section 2 provides the background of the present study. Section 3 describes the proposed approach in some details while the postprocessing operations are described in Sect. 4. Section 5 presents extensive simulation results using the sample scene images of ICDAR 2003 Robust Reading Competitions Database [7]. Finally, Sect. 6 concludes the article.

2   Background

According to the dichromatic reflection model [8], most of inhomogeneous objects have two kinds of reflections: diffuse and specular. In the dichromatic reflection model, the spectral factor is a weighted sum of the specular and diffuse reflectance functions. If we use a digital camera with three sensors (RGB) to capture an image, the color I at each pixel x can be described as a linear combination of two components D and S as follows:

$$I(x) = m_d(x)D(x) + m_s(x)S(x),$$

where $I(x) = [I_r(x), I_g(x), I_b(x)]^T$ is the RGB color at pixel x in the captured image. $D(x) = [D_r(x), D_g(x), D_b(x)]^T$ and $S(x) = [S_r(x), S_g(x), S_b(x)]^T$ denote the diffuse and specular chromaticity at pixel x, respectively. $m_d(x)$ and $m_s(x)$ are two factors of the two reflections which depend on the local geometry of the pixel on the surface. The first step of the proposed approach identifies the specular pixels of input image. Once the specular pixels are identified, regions around it are selected and different enhancement techniques are applied on these specular parts. Enhancement techniques involve estimation of the foreground pixels by applying Otsu's method separately at each such region. Enhancement of the remaining parts of the image is considered separately. Thus what we obtain is a more or less binarized image containing the probable text regions along with possibly a huge amount of noise as well. So, in the next step, we remove those noisy pixels and enhance the text regions further as detailed in the following Sections.

3   Proposed Methodology

3.1   Separation of Specular and Diffuse Pixels

Yuan He et al. [9] proposed a fast and efficient approach to distinguish between specular and diffused pixels. We use the same approach. The chromaticity σ of each pixel is defined as

$$\sigma(x) = \frac{I(x)}{I_r(x) + I_g(x) + I_b(x)},$$

where $\sigma = [\sigma_r, \sigma_g, \sigma_b]^T$. The maximum chromaticity is defined as $\hat{\sigma} = \max(\sigma_r, \sigma_g, \sigma_b)$. According to the mechanism, we can get the total diffuse intensity of specular pixels as

$$\sum_{i \in \{r,g,b\}} I_i^{diff} = \frac{\hat{I}(x)[3\hat{\sigma}(x) - 1]}{\hat{\sigma}(x)[3\hat{\lambda}(x) - 1]}$$

where $\hat{I} = \max(I_r, I_g, I_b)$ and $\hat{\lambda} = \max(\lambda_r, \lambda_g, \lambda_b)$ is the maximum diffuse chromaticity which is set to an arbitrary value ($\frac{1}{3} < \hat{\lambda} \le 1$). The specular component is obtained as

$$I^{spec}(x) = \frac{\sum I_i(x) - \sum I_i^{diff}(x)}{3}.$$

Then, the diffuse component of the specular pixels is achieved by subtracting the intensity with the specular component $I^{diff}(x) = I(x) - I^{spec}(x)$. A pixel x is labeled as specular if $I^{diff}(x) > \tau$, a constant threshold, otherwise it is

(a)   (b)

Fig. 2. (a) Identified specular pixels of the sample image of Fig. 1(b), (b) final output of the proposed method for binarization of specular region corresponding to the sample image of Fig. 1(b).

labeled as diffuse. However, in our case, choosing a very high value of τ enables us to identify only those pixels which are highly specular, that is information contained in those pixels can be assumed to be lost under the high flash of light. Hence unlike [9], we don't need to enhance the specular pixels as it doesn't contain any information. This assumption has the drawback that very bright (almost white) text pixels get identified as specular pixels. Figure 2(a) shows the identified specular pixels of the sample image in Fig. 1(b). However, usually, the boundary pixels of such white text components are not very bright unlike its interior pixels and so the information is not completely lost as it can be seen from the output of the proposed method shown in Fig. 2(b). On the other hand, if the background of non-white text is very bright (almost white), although the region get selected as specular but when the same is binarized, the foreground text information is preserved. In fact, if the color of the text is darker than the background, the proposed method can efficiently extract the text as it may be seen from some of the results shown in Sect. 5.

3.2   Binarization of Specular Regions

Once the specular pixels are identified, connected region(s) of these pixels are obtained to apply Otsu's binarization method separately on each of them. Figure 3(a) shows the identified specular pixels (white) of the sample image in Fig. 1(a). Smallest enclosing rectangles around each of the connected components of these specular pixels are constructed and these are shown in Fig. 3(b). Overlapping rectangles or rectangles separated by a small distance (say, less than 100) are merged together as shown in Fig. 3(c). In the next step, we consider all such rectangles barring possibly a few very small ones (less than 100 pixels) treated as common noise. Let's denote such a rectangular region by R. We apply well-known Otsu's binarization method on R and obtain the corresponding binarized region R' as shown in Fig. 3(d) and this binarized result resembles the results in Fig. 2(b).

(a)   (b)   (c)   (d)

Fig. 3. Binarization of the specular regions: (a) specular pixels of the sample image in Fig. 1(a), (b) smallest enclosing rectangles of connected components of specular pixels shown in (a), (c) after merging of rectangles shown in (b), (d) after binarization of each such rectangle.

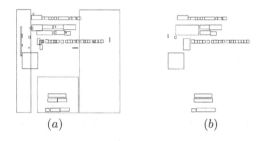

(a)   (b)

Fig. 4. (a) Bounding rectangles of connected edge components obtained from the image of Fig. 1(a), (b) candidate text regions of the sample of Fig. 1(a).

3.3   Binarization of the Whole Input Image

In this step, we consider the whole gray-level image. We first consider reduction of noise by applying Gaussian smoothing with 5×5 kernel for a few successive times. Next, we apply Canny's method [10] and the distinct connected components of the resulting edge image are identified. Bounding rectangles of such connected components obtained from the input image sample of Fig. 1(a) are shown in Fig. 4(a). We consider a set of empirical rules used in previous studies [11,12] such as (i) aspect ratio of a text region lies between 0.1 and 10, (ii) height of a text region is larger than 10 pixels, (iii) both height and width of a text region cannot be larger than half of the corresponding size of the input image etc. to filter out majority of non-text components and two or more among the remaining rectangular regions are merged whenever these overlap. The resulting rectangles provide the candidate text regions of the input image. These regions corresponding to the sample image of Fig. 1(a) are shown in Fig. 4(b).

Now, Otsu's thresholding method is applied individually on each of the above rectangular candidate text regions and respective threshold values are obtained. Pixels contained in a rectangular region are labelled 'BLACK' or 'WHITE' based on its threshold value. The result of applying Otsu's thresholding method

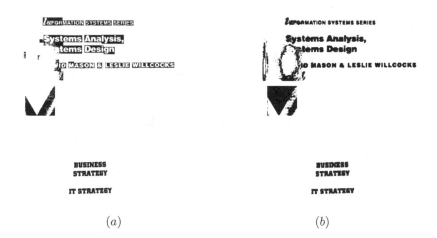

(a) (b)

Fig. 5. (a) Result of binarization by applying Otsu's method separately on each candidate text region, (b) the binarized image of the sample in Fig. 1(a) after the pore filling step.

locally on the sample image in Fig. 4(a) is shown in Fig. 5(a). Here, it may be noted that the above procedure may assign either 'BLACK' or 'WHITE' label to a text component. So, next we consider the following procedure to ensure that text components always get the label 'BLACK' and background pixels always get 'WHITE' label. A similar approach was considered before in [13].

We consider the labels of following 12 pixels consisting of 3 pixels of each of four corners of a binarized candidate text region B, i.e., the labels of

$$\{B(x, y), B(x, y + h - 1), B(x + w - 1, y), B(x + w - 1, y + h - 1),$$
$$B(x + 1, y), B(x + 1, y + h - 1), B(x + w - 2, y), B(x + w - 2, y + h - 1),$$
$$B(x, y + 1), B(x, y + h - 2), B(x + w - 1, y + 1), B(x + w - 1, y + h - 2)\},$$

where w and h are respectively the width and height of B and $B(x, y)$ is its top-left pixel. We obtain the count of 'WHITE' labels among these 12 pixels and if this count is less than 6, then we interchange the labels of each pixel of the candidate text region B, otherwise these are left unchanged. Finally, the specular regions R (identified in Sect. 3.2) of the above binarized image are replaced by the corresponding binarized specular regions R'.

4   Postprocessing

Since the binarized image still contains significant amount of noise, in this Section, we describe a sequence of postprocessing operations to produce improved binarized image consisting of fewer noise components and cleaner texts. The first postprocessing step is filling of pores which is followed by removal of components of non-uniform thickness or components consisting of only a few pixels.

4.1 Filling of Pores

Usually, binarized components extracted from glare-affected regions have many pores as it may be verified with the three characters ('Sys') of Fig. 3(d). It often causes difficulties towards the success of postprocessing operations. So, here we first consider filling of similar pores on the binarized image (say, B) obtained in the above Sect. 3 as described below.

At each background pixel of B we consider a 3×3 window and count the number of foreground pixels inside this window. If this number exceeds two-third of the total number of pixels inside the window (i.e., 7 or more), then the background pixel is converted to a foreground pixel. This technique successfully fills majority of the pores that appear on the foreground components of the binarized image as shown in Fig. 5(b). This output image is next subjected to the Euclidean Distance Transform as follows.

4.2 Identification of Components of Non-uniform Thickness

It has already been discussed in the literature [11,12,14] that an intriguing property of text components is that these have more or less uniform thickness compared to their non-text counterparts. The well-known Euclidean distance transform (DT) [15] has been used before [11,12] to estimate the thickness of a connected foreground component of binary image. Each pixel of the foreground component in the transformed image (denoted as $D^\star$) is set to a value equal to its distance from the nearest background pixel. This has been explained with the help of Fig. 6. In Fig. 6(a), '1' represents foreground pixels and '0' represents background pixels. Each object pixel of this binary image is assigned a value which is its distance from the nearest background pixel as shown in Fig. 6(b). At each non-zero pixel of D^*, we consider a 3×3 window to obtain the local maximum at that pixel. If this pixel value equals the local maximum, we store the pixel value in a list $< T >$ for further processing. In fact, such a pixel value (a local maximum value) is an estimate of half of the local stroke thickness. Finally, we compute the mean (μ) and the standard deviation (σ) of the local stroke thickness values stored in $< T >$. If $\mu > 2\sigma$, (well-known 2σ limit used in statistical process control), we decide that the thickness of the underlying stroke is nearly uniform and select the foreground component as a candidate text component. The above operation helps to filter out majority of non-text components present in the output image produced by the method of Sect. 3.3 as shown in Fig. 6(c).

We applied DT operation after filling of pores as described in Sect. 4.1 because otherwise some of the text components would get removed by this operation as shown in Fig. 6(d). On the other hand, it can be seen from the result shown in Fig. 6(c), the individual characters are so thickened due to pore filling operation that adjacent characters often get merged. Since joining of adjacent characters may pose problem to the OCR software, we next replace the bounding rectangles of connected components in the latest output image by similar rectangles taken from the output of the binarization module described in Sect. 3.3. The result of this replacement operation is shown in Fig. 6(e).

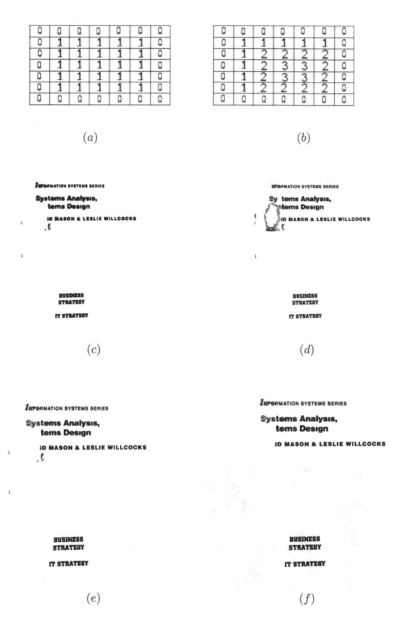

Fig. 6. (a) A binary image, (b) distance transform of the image in (a), (c) after removal of non-text components of the image in Fig. 5(b) with the help of DT, (d) result of DT operation without filling of pores – the first character 'i' in the first row and third character 's' in the second row have been removed and a non-text component couldn't be removed, (e) after replacement of thick foreground components by the respective components before pore filling operation, (f) small foreground components consisting of 100 or fewer pixels are removed.

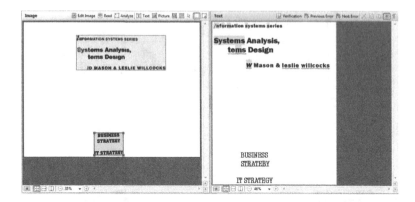

Fig. 7. Screenshot of the FineReader software output when fed with the final output shown in Fig. 6(f). Porus characters 'Sys' at the beginning of the 2nd line have been correctly recognized by the software.

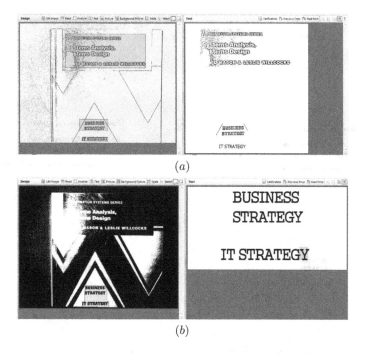

(a)

(b)

Fig. 8. Screenshots of the FineReader 11 Software output when it is fed with the scene image shown in Fig. 1(a) after binarization by (a) adaptive thresholding, (b) Otsu's thresholding methods.

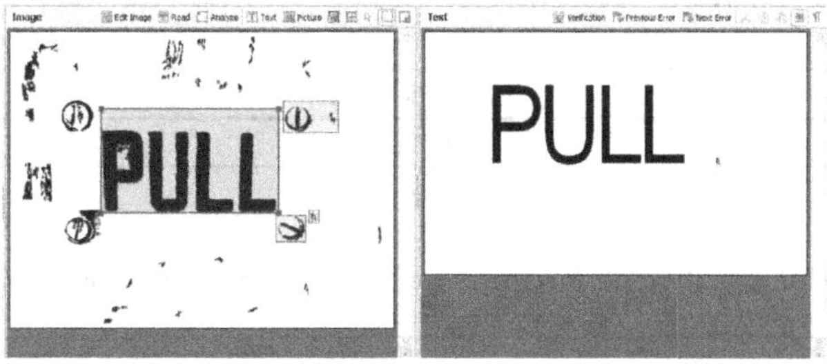

(*a*)

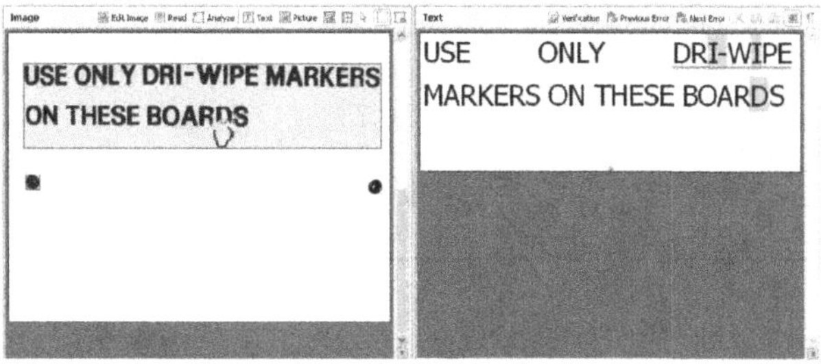

(*b*)

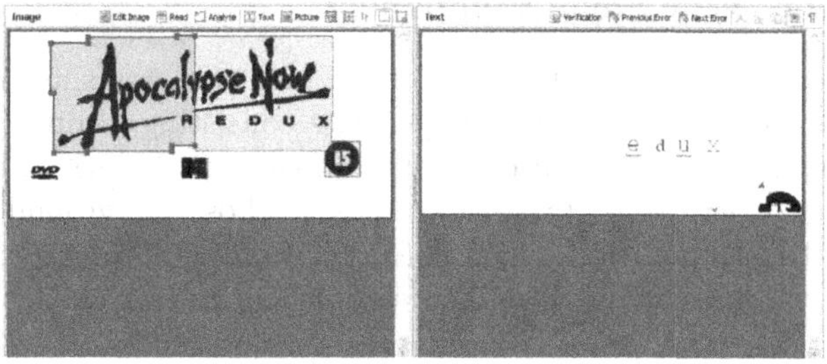

(*c*)

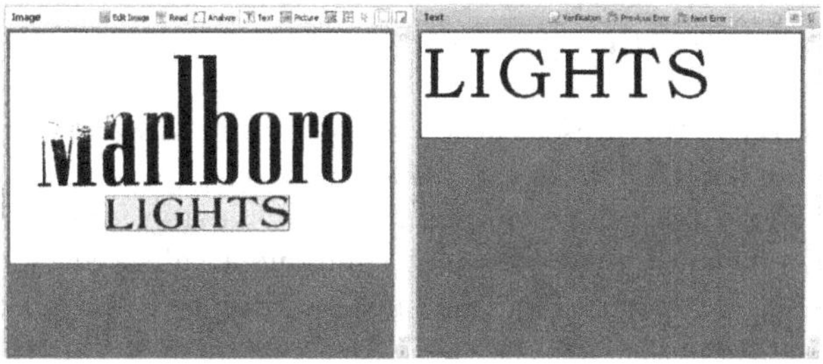

(*d*)

Fig. 9. The input to the OCR software, i.e., the output of proposed strategy consists of (*a*) several non-text components but the OCR software removed all of them barring only one, (*b*) only 3 non-text components and the OCR software removed each of them, (*c*) two artistically written words and another word 'REDUX' with its characters widely spaced and OCR software didn't recognize both the artistic words as texts, (*d*) two words 'Marlboro' and 'LIGHTS', the middle two characters of 'Marlboro' are sufficiently larger than others and the OCR software didn't detect this word as text while it correctly recognized the other word written in plain style.

Table 1. OCR based results based on images with only plain texts

Algorithm	Precision(p)	Recall(r)
Proposed method	0.79	0.77

Table 2. Comparative performance analysis of text detection approaches

Algorithm	Precision(p)	Recall(r)
Roy Chowdhuri et al. [12]	0.57	0.59
Kasar et al. [13]	0.63	0.59
Epshtein et al. [14]	0.73	0.60
Chen et al. [16]	0.73	0.60
Merino-Gracia et al. [17]	0.51	0.67
Zhang and Kasturi [18]	0.67	0.46
Proposed method	0.63	0.65

4.3   Removal of Very Small Components

As it can be seen from the latest output of the previous step shown in Fig. 6(e), it still contains several small noise components consisting of only a few pixels. On the basis on extensive simulation results on the training set of ICDAR 2003 robust reading competition database, we remove those foreground components of the output of last operation which consists of only 100 or fewer pixels. The final output of the present work on the sample image in Fig. 1(a) is shown in Fig. 6(f).

5   Simulation Results

A broad motivation of the present study is to binarize a scene image efficiently such that available OCR software can recognize texts in them. The OCR output shown in Fig. 7 shows that the proposed method is a step forward in the above direction. It can efficiently deal with text components affected by glare. It may also be noted that the porus text 'Sys' in the binarized image do not pose a problem to the state-of-the-art software. Also, we have shown the OCR software outputs on the same scene image when the well-known adaptive method or Otsu's global method were used for binarization. In Fig. 8(a) (produced by adaptive method), only the last line of texts 'IT STRATEGY' could be recognized and in Fig. 8(b) (produced by Otsu's global method), only the lower block of texts containing three lines 'BUSINESS', 'STRATEGY' and 'IT STRATEGY' could be recognized. Thus, in both cases the upper block of text could not be identified as texts although when the binarized image produced by the proposed method was fed to the OCR software, it identified both upper and lower blocks of texts as shown in Fig. 7.

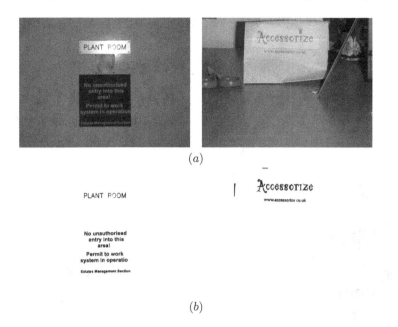

(a)

(b)

Fig. 10. (a) Two image samples having varying contrasts, (b) simulation results of the proposed method on the image samples shown in (a).

We implemented the proposed approach using DEV-C++ and OpenCV library under Windows environment. For evaluation purposes, we used a similar strategy as that of the ICDAR2003 robust reading competition [7] and also several other studies [12–14, 16–18]. We obtained the two measures precision (p) and recall (r) of our extraction results. Expressions used for computation of p and r values are as follows.

$$p = \frac{P \cap T}{P} \text{ and } r = \frac{P \cap T}{T},$$

where P is the area of the minimum enclosing rectangles of all connected components extracted by the proposed strategy and T is the area of all the rectangles provided as the ground truth together with the sample images.

Since the motivation of the present study is to process an image suitably before it is being fed to an OCR software for text recognition, an effective evaluation strategy would be to obtain the value of P from the OCR software output. In this context, it may be noted that this strategy has both positive and negative aspects. The positive aspect is that a state-of-the-art OCR software can efficiently filter out several artifacts retained in the output image of the text extraction method as shown in Fig. 9(a), (b) while the negative aspect is that an existing OCR fails to recognize texts of artistic styles as shown in Fig. 9(c), (d). So, we computed values of p and r based on results of FineReader 11 software

Fig. 11. A few image samples of ICDAR 2003 Robust Reading Competition Database are shown in the left column. Each of these sample images have multi-oriented texts. Results of the proposed method on these sample images are shown in the right column

Fig. 12. (a) & (c) A few samples from ICDAR 2003 robust reading competition database affected by non-uniform lighting or shadow, (b) & (d) results of the proposed method on respective image samples shown in (a) & (b).

and on only those test samples which do not have any text of similar artistic styles. These results are shown in Table 1.

However, for the purpose of fair comparison with the existing results available in the literature, we also computed the values of p and r directly based on the output of our proposed approach before feeding them to an OCR software and the entire set of test samples of ICDAR 2003 robust reading competition database. The comparative results are listed in Table 2. It shows that the performance of the proposed strategy is comparable with the state-of-the art results.

(a)   (b)   (c)

Fig. 13. (a) A sample image of ICDAR 2003 database, (b) the binarized image before restoration of small components in the close proximity of a text component, (c) the binarized image after restoration of smaller text components.

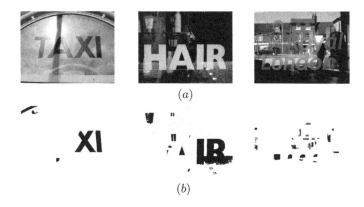

Fig. 14. (a) Three image samples with texts on transparent glass doors, (b) respective outputs of the proposed method.

On the other hand, the proposed approach is a generic one in the sense that it can extract texts of arbitrary orientations and also it can partially recover from degradation affected by specular reflection. Outputs of the proposed approach on a few images of different difficulties are shown in Figs. 10, 11 12.

6   Conclusions

The present study shows that although Otsu's global thresholding technique is not effective on scene images but the same provides superior performance when it is applied locally on smaller rectangular regions enclosing connected components of the Canny edge image of such a scene image. The proposed approach can partially recover distortions due to specular reflections. On the other hand, our aggressive noise cleaning approach removing smaller components consisting of 100 or lesser pixels may often remove certain punctuation marks and the dots of 'i', 'j'. It is true that the removal of dots of 'i' or 'j' may not be a serious issue since such a loss is usually taken care of by the OCR software as it can be seen from

Fig. 7. However, the loss of punctuation marks is required to be handled at the binarization module and the same may be achieved by considering a rectangle around each connected text component of height and width 5 % larger than those of the respective text components and if there exists a non-text component inside this rectangle and removed earlier due to its small size, then the same is restored. In Fig. 13 we have shown restoration of such small text components as in the above. Although the proposed method can efficiently extract texts from various types of outdoor scene images, but its performance towards texts written on transparent glass doors is miserable as it can be seen from Fig. 14.

References

1. Otsu, N.: A threshold selection method from gray-level histograms. IEEE Trans. Syst., Man Cybern. **9**(1), 62–66 (1979)
2. Kittler, J., Illingworth, J., Foglein, J.: Threshold selection based on a simple image statistic. Comp. Vision Graph. Image Proc. **30**(2), 125–147 (1985)
3. Sauvola, J.J., Pietikainen, M.: Adaptive document image binarization. Patt. Recog. **33**(2), 225–236 (2000)
4. Niblack, W.: An Introduction to Digital Image Processing. Prentice Hall, New York (1986)
5. Stathis, P., Kavallieratou, E., Papamarkos, N.: An evaluation technique for binarization algorithms. J. Univ. Comp. Sci. **14**(18), 3011–3030 (2008)
6. Peng, X., Setlur, S., Govindaraju, V., Sitaram, R.: Markov random field based binarization for hand-held devices captured document images. In: Proceedings of Indian Conference on Comp. Vision Graph. Image Proceedings, pp. 71–76 (2010)
7. Lucas, S.M., Panaretos, A., Sosa, L., Tang, A., Wong, S., Young, R.: ICDAR 2003 robust reading competitions. In: Proceedings of the 7th Internationl Conference on Document Analysis and Recognition, pp. 682–687 (2003)
8. Shafer, S.A.: Using color to separate reflection components. Color Res. Appl. **10**, 210–218 (1985)
9. He, Y., et al.: Enhancement of camera-based whiteboard images. In: XVII-DRR (SPIE Proceedings Series, vol. 7534, pp. 1–10 (2010)
10. Canny, J.: A computational approach to edge detection. IEEE Trans. Patt. Anal. Mach. Intell. **8**(6), 679–698 (1986)
11. Roy Chowdhury, A., Bhattacharya, U., Parui, S.K.: Text detection of two major Indian scripts in natural scene images. In: Iwamura, M., Shafait, F. (eds.) CBDAR 2011. LNCS, vol. 7139, pp. 42–57. Springer, Heidelberg (2012)
12. Roy Chowdhury, A., Bhattacharya, U., Parui, S.K.: Scene text detection using sparse stroke information and MLP. In: Proceedings of International Conference on Pattern Recognition, pp. 294–297 (2012)
13. Kasar, T. et al.: Font and background color independent text binarization. In: Proceedings of CBDAR, pp. 3–9 (2007)
14. Epshtein, B., Ofek, E., Wexler, Y.: Detecting text in natural scenes with stroke width transform. In: Proceedings of CVPR, pp. 2963–2970 (2010)
15. Borgefors, G.: Distance transformations in digital images. Comp. Vis. Graph. Image Proc. **34**, 344–371 (1986)
16. Chen, H., et al.: Robust text detection in natural images with edge-enhanced maximally stable extremal regions. In: Proceedings of ICIP (2011)

17. Merino-Gracia, C., Lenc, K., Mirmehdi, M: A head-mounted device for recognizing text in natural scenes. In: Proceedings of CBDAR, pp. 27–32 (2011)
18. Zhang, J., Kasturi, R.: Text detection using edge gradient and graph spectrum. In: Proceedings of ICPR, pp. 3979–3982 (2010)

Scene Text Detection via Integrated Discrimination of Component Appearance and Consensus

Qixiang Ye$^{(\boxtimes)}$ and David Doermann

Institute of Advanced Computer Studies, University of Maryland, College Park, USA
{qxye,Doermann}@umiacs.umd.edu

Abstract. In this paper, we propose an approach to scene text detection that leverages both the appearance and consensus of connected components. A component appearance is modeled with an SVM based dictionary classifier and the component consensus is represented with color and spatial layout features. Responses of the dictionary classifier are integrated with the consensus features into a discriminative model, where the importance of features is determined with a text level training procedure. In text detection, hypotheses are generated on component pairs and an iterative extension procedure is used to aggregate hypotheses into text objects. In the detection procedure, the discriminative model is used to perform classification as well as control the extension. Experiments show that the proposed approach reaches the state of the art in both detection accuracy and computational efficiency, and in particularly, it performs best when dealing with low-resolution text in clutter backgrounds.

Keywords: Text detection · Component · Discrimination

1 Introduction

Text detection and recognition in natural scene images has recently received increased attention of the computer vision community [1–3]. There are at least three reasons for this trend. First is the demand for applications to read text for indexing, especially on mobile devices and in streetview data. Compared with the other image objects, text is embedded into scenes by humans, typically with the intention that it be read. Second is the increasing availability of high performance mobile devices, which creates an opportunity for imagery acquisition and processing anytime, anywhere and makes it convenient to recognize text in various environments. The third is the advance in computer vision technologies, which is making it feasible to address these more challenging problems.

As an important prerequisite for text recognition, text detection in natural scene images still remains an open problem due to factors including complex background, low quality images, variation of text content and deformation of text appearance.

M. Iwamura and F. Shafait (Eds.): CBDAR 2013, LNCS 8357, pp. 47–59, 2014.
DOI: 10.1007/978-3-319-05167-3_4, © Springer International Publishing Switzerland 2014

There are generally two classes of methods used in existing scene text detection approaches: connected components[1] based and sliding window classification based.

The component based methods often use color [3], point [4], edge/gradient [5], stroke [6,7], and/or region [8–11] features or a hybrid of them [12,13] to localize text components, which are then aggregated into candidate text regions for further processing. Recently, Maximally Stable Extremal Regions (MSERs) based text detection has been widely explored [8–11]. The main advantage of these approaches over other component based approaches is rooted in the effectiveness of using MSERs as character/component candidates. It is based on the observation that text components usually have higher color contrast with their backgrounds and tend to be form homogenous color regions. The MSER algorithm adaptively detects stable color regions and provides a good solution to localize the components.

In [8,10], MSERs from H, S, I and gradient channels are integrated to detect components. An exhaustive search is then applied to group components into regions and a text level classifier is used for classification of these regions. In [11], Koo et al present a text detection approach based on MSERs and two classifiers. The first classifier is trained on AdaBoost that determines the adjacency relationship and cluster components by using pairwise relations. The second classifier is a multi-layer perceptron classifier that performs text/non-text classification on normalized candidate regions. Benefitting from the learning method for clustering components, their approach won the ICDAR2011 Robust Reading Competition [14].

Although existing MSER based approaches report promising performance, problems remain. In particular, the approaches detect a large number of candidate components and neither effective classification nor component grouping has been adequately addressed. Existing rule based approaches generally require fine-tuned parameters. Clustering based methods require well defined/learned criterion to determine the number of clusters and when to stop the clustering.

At the same time, sliding window methods usually train discriminative models to detect text with a multi-scale sliding window classification [15–18]. The advantage of this kind of method lies in the fact that the training-detection architecture is simpler and more discriminative than component based approaches. Disadvantages lie in that sliding window classification at multiple scales is often time-consuming and computationally expensive. It is also difficult to detect non-horizontal text because a tremendous large number of windows that would need to be considered in a three dimensional space of scale, position and orientation. After text components/patches are localized, existing methods usually use grouping-and-classification strategies to perform text level processing. Detected patches can be grouped into text regions with morphological operations [15], Conditional Random Field [16] or graph methods [17,18]. In existing approaches, grouping and classification procedures are often separated where consensus among components/patches is used for grouping and appearance is

---

[1] "Connected component" is shorted as "component" in the followings.

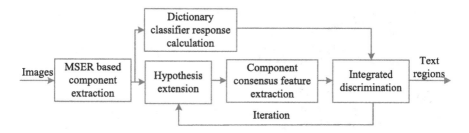

Fig. 1. Block diagram of the proposed text detection approach.

used for classification. In our work, we show that consensus among components can also be used for text/non-text classification. A group of components of less consensus on color and spatial alignment are less likely to be a text objects while a "good" component grouping strategy should benefit the following classification procedure.

In this paper, we propose a new approach to detect and localize scene text by integrating both appearance and consensus representation of components into a discriminative model. The component appearance representation is built on Histogram of Oriented Gradient (HOG) features and a sequence of SVMs to build a dictionary classifier. The consensus representation includes color distance, color variance and spatial distance of components. The functions of the discriminative model are twofold: classifying text/non-text and determining whether components should be grouped. When performing text detection, MSERs are first extracted as candidate components. Text hypotheses are then generated on MSER pairs. The hypotheses are extended iteratively until the output of the discriminative model is negative. Block diagram of the proposed approach is shown in Fig. 1.

The remainder of this paper is organized as follows. The text detection approach is described in Sect. 2. Experiments are provided in Sect. 3 and conclusiolns in Sect. 4.

2 Text Detection Approach

The proposed text detection approach includes the following procedures: MSER based component extraction, training of the component dictionary classifier, integrated discriminative model and a text detection algorithm.

2.1 Component Extraction

Among a number of component extraction methods, we have adopted the MSER algorithm because it shows good performance with a linear computation cost [19]. This algorithm finds local shapes that are stable over a range of thresholds, allowing us to find most of the text components [11]. In each channel, the MSER algorithm yields components that are either darker or brighter than their

Fig. 2. MSER bounding boxes from three channels. (a) Luminance channel, (b) chrominance channel, (c) gradient channel and (d) detected text objects.

surroundings. We use a Gamma correction ($\gamma = 1.0$) on the image as a pre-processing step so that low contrast text components can be correctly localized. MSERs from the luminance, chrominance and gradient channels are extracted and pooled. In Fig. 2, we illustrate bounding boxes of darker components. Note that some components overlap with each other, showing the complex spatial relationships among them.

2.2   Component Dictionary Classifier

When detecting text in the natural scenes, we need to effectively discriminate text from other objects. Researchers have tried to model text as "structured edges", "a series of uniform color regions", "a group of strokes", or "texture" or hybrid of these. However, there are many objects such as leaves or window curtains that have similar edge, stroke or texture properties as text, making it difficult to find effective features and methods to discriminate text from these objects. In this paper we propose a more precise definition of "text patterns" using a sequence of classifiers corresponding to components (characters or groups of them). It is based on the fact that character patterns have been well defined and people have explored many effective features to represent characters.

HOG features, a kind of state-of-the-art features for object representation, are employed as component representation. As shown in Fig. 2(a), when extracting

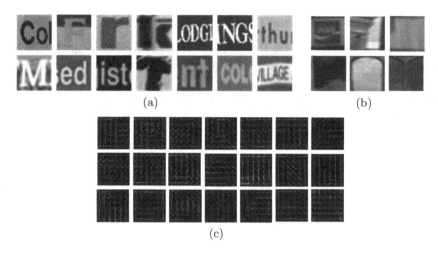

(a)   (b)

(c)

Fig. 3. Illustration of component dictionary classifier. (a) component samples, (b) negatives and (c) visulization of 21 normal vectors of the dictionary classifier, in which 20 normal vectors are for components and the last one is for negtive samples.

HOG features, a 28×28 training sample is divided into cells of size 4×4 pixels, and each group of 2×2 cells is integrated into a block using a sliding window, and blocks overlap with each other. We first calculate the gradient orientation of each pixel. In each cell, we calculate nine-dimensional HOG features by calculating the nine-bin histogram of gradient orientations of all pixels in this cell. Each block contains four cells, on which 36-dimensional features are extracted. Each sample is represented by 36 blocks, on which 1296-dimensional HOG features are extracted. Component samples are partitioned into K groups with a K-means clustering algorithm in the HOG feature space. Considering the difficulty when clustering samples in the high dimensional feature space, we use the methods proposed in [20] to improve the clustering results, iteratively.

Clustered training samples of components and their negative images are shown in Figs. 3(a) and (b) respectively. It can be seen that the component samples include single characters or several characters (or character parts) extracted by the MSER algorithm. Samples of an aspect ratio larger than 5.0 are considered to be seriously touching and are discarded before training. The $(K + 1)$th group corresponds to the negative samples. A multi-class SVM training algorithm is used to train the dictionary classifier that containing K linear SVMs $f_k(x) = w_k^T \cdot x + b_k, k = 1, ...K$ that correspond to K component clusters. For the multi-class training, an one-against-all strategy is adopted.

When performing classification, the output of a feature vector x from the dictionary classifier is the maximum response of K linear SVMs as

$$f(x) = \arg\max \{f_k(x)\} \tag{1}$$

2.3   Component Consensus Feature Extraction

Component consensus includes the pairwise relations of components and the holistic variance of grouped components. Let i and j denote the indexes of two components.

Given a text $X = (x_1, ..., x_i)$ with the last component x_i and an isolated component x_j, the spatial relations between text X and component x_j is described with five features as follows:

– Color difference feature

$$\varphi_1 (X, x_j) = \varphi_1 (x_i, x_j) = \|c_i - c_j\|_2 \tag{2}$$

where c_i and c_j are mean color of component i and j.
– Spatial distance features (symbols are defined in Fig. 4)

$$\varphi_2 (X, x_j) = \varphi_2 (x_i, x_j) = \frac{v_{ij}}{\min (h_i, h_j)} \tag{3}$$

$$\varphi_3 (X, x_j) = \varphi_3 (x_i, x_j) = \frac{d_{ij}}{\min (h_i, h_j)} \tag{4}$$

– Alignment features (symbols are defined in Fig. 4)

$$\varphi_4 (X, x_j) = \varphi_4 (x_i, x_j) = \frac{|h_i - h_j|}{\min (h_i, h_j)} \tag{5}$$

$$\varphi_5 (X, x_j) = \varphi_5 (x_i, x_j) = \frac{o_{ij}}{\min (h_i, h_j)} \tag{6}$$

Assuming that x_i is merged into text X and forms a new text, then we can calculate the variance of color mean values of components, as follows:

$$\varphi_6 (X, x_j) = \varphi_6 (X \cup x_j) = Variance (c_1, c_2, ..., c_i, c_j) \tag{7}$$

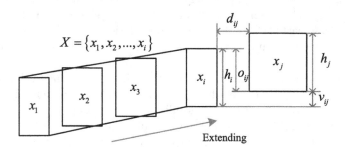

Fig. 4. Illustration of spatial relationship of component x_i and x_j.

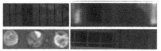

Fig. 5. Illustration of text (left)and mined hard negative samples (right).

2.4   Integrated Discriminative Model

Using the dictionary classifier defined in Sect. 2.2 and consensus feature extraction procedure in Sect. 2.3, the text discriminative model is defined as

$$F(\tilde{X}) = F(X \cup \tilde{x}) = W^T \cdot \begin{pmatrix} \psi(X) \\ \varphi(X, \tilde{x}) \end{pmatrix} - B \qquad (8)$$

where $\psi(X)$ denotes response feature extraction from the dictionary classifier, which outputs two features: the average response $\frac{1}{|X|} \sum_{x_n \in X} f(x_n)$ and the maximum response of all the components in X as $\max(f(x_n)), n = 1, ..., |X|$. $\varphi(X, \tilde{x})$ denotes the component consensus feature extraction which includes the six dimensional features defined by (2)–(7). $\tilde{x}$ is a component being considered for inclusion into X. W^T is a weight vector related to importance of each dimension of features and B is a threshold for the model when performing text classification. If $F(\tilde{X})$ outputs a positive value it means that $\tilde{X}$ is a text object; otherwise a non-text object.

Given the text samples and their negatives, we calculate dictionary classifier responses of components and component consensus features. The features are input into a linear SVM to train the integrated discriminative model of (8). To obtain a high performance discriminative model it is often important to use large training sets. In the case of text there is few text samples for training but a lot of negatives from background. Therefore a Bootstrapping procedure is adopted to train a model with an initial subset of random negative examples, and then collect negative examples that are incorrectly classified by this initial model to form a set of hard negatives, as shown in Fig. 5. A new model is trained with the hard negative examples and the process is repeated.

2.5   Text Detection

Text detection is a procedure of hypothesis-generation and hypothesis-extension. Hypotheses are generated from component pairs. As the MSER algorithm can generate thousands of component candidates in images of complex backgrounds, using all of the component pairs to generate text hypotheses is time consuming. Two loose constraints on the component spatial distance and alignment are used to reduce number of hypothesis. We then extend the hypotheses, as illustrated by Fig. 4, iteratively to obtain text objects. The text detection procedure is described as the following algorithm.

Algorithm 1. Text detection algorithm

---

1 **Hypothesis generation**
  On extracted components, generate text hypotheses from component pairs that feeds the following loose constrains:
  1.1 Two spatial distances of component x_i, x_j, defined in (3) and (4), are less than 1.0.
  1.2 The vertical overlap of two components, defined in (6), is larger than 0.5.
2 **Hypothesis extension**
  2.1 Randomly select a text hypothesis X;
  2.2 Find the component set $\{x_j\}, j = 1, ..., J$ that satisfies constrains 1.1 and 1.2.
  2.3 Select the nearest component to extend by

$$\arg\min_{x_j} \left(\varphi_1\left(X, x_j\right) \cdot \varphi_2\left(X, x_j\right) \cdot \varphi_3\left(X, x_j\right) \right), j = 1, ..., J. \quad (9)$$

  2.4 Classify the extended object $\tilde{X} = X \cup x_j$ with (8). If $\tilde{X}$ is classified as text, $X \leftarrow \tilde{X}$, remove component x_i from the component set and goto step 2.3; otherwise goto 2.1.
  2.5 If there is no hypothesis that can be extended, goto step 3; otherwise goto 2.1.
3 **Merging overlapped text**
  Merge text objects that overlap with each other and outputs their bounding boxes and fitted border lines.

---

3   Experiments

3.1   Datasets

We use two different datasets, ICDAR 2011 scene text dataset [15] and the Street View Text (SVT) dataset [16] to perform evaluation. The ICDAR 2011 dataset is widely used for benchmarking scene text detection algorithms. Most of the text objects in it are captured at short distances and the main challenges are from the large scale variance and uneven illumination. The SVT dataset contains text objects from Google street video frames and most of the text objects are captured at middle distances. The main challenge is from the degradation of image quality and the complex background.

The ICDAR2011 dataset contains 849 training text samples from 229 images and 1190 test samples from 255 images. The SVT dataset contains 257 training samples from 100 images and 647 test text samples from 250 images. We extracted 9930 components and 1620 text samples for training. We also mined 6000 negative component samples and 1000 negative text samples.

On the ICDAR2011 dataset, our evaluation protocol is consistent with the ICDAR2011 robust reading competition [14] (using the Deteval software). Precision, recall and a harmonic mean are used as the protocol to perform evaluation. On the SVT dataset, the bounding boxes in groundtruth are not precise. In such a case, precision is defined as the ratio between area of intersection regions and that of detected text regions, recall is obtained from the ratio between area of intersection regions and that of ground truth regions.

3.2   Effects of Parameters

We have conducted experiments to show the effects of different color channels as shown in Table 1. With L, U and V color channels the precision is four percent higher than in R, G and B channels. More than three point Harmonic mean performance improvement is observed when using L, U, V and the gradient channels compared with the R, G and B channels. This shows the effectiveness of the combination of multiple channels when performing detection.

We use the harmonic mean rate as the criterion to determine the positive component classifier number K in the dictionary classifier, as shown in Fig. 6. In the experiments, it is found that the more the training samples, the larger the value of K is. For the current training set, a scope of [24–32] of K reports the best performance.

We have also illustrated precision-recall curves for different classification threshold T in Fig. 7. Specifically, Fig. 7 shows that the precision could significantly drop when the recall rate increase up to 65 %. It can be calculated that when setting T=0.2 and T=0.4, we can obtain the best performance on ICDAR 2011 and SVT datasets, respectively.

3.3   Results and Comparisons

In Table 2 we compare our approach with other state-of-the-art approaches on the ICDAR2011 dataset. Compared with the competition winner, our proposed approach has improvement on both the precision and recall rates. In particularly, our method can keep a higher precision rate without significant recall drop. In Table 3 we compare our proposed approach with two published state-of-the-art approaches [10,16] on the SVT dataset. It can be seen that our approach shows significant improvement on the Harmonic mean (more than 12 %). Nevertheless, on this dataset, all of the compared approaches reports low recall and precision rates. The main reason is for the image quality degradation after the video compression and decompression procedures. When performing detection, our approach runs at a speed about 1.1 images per second (for images of width 756 pixels) on a PC with an Intel CORE i5 CPU. It is observed in experiments that the speed mainly depends on the MSER parameter Δ that represents the range of intensities where the regions are stable [19]. When setting $\Delta = 2$, thousands of components can be detected and the detection speed drops to 0.45 images/second. When setting $\Delta = 6$ the number of components reduces

Table 1. Performance (%) comparison of text detection different color channels.

Color channels	Recall	Precision	Harmonic mean
LUV+Gradient	64.64	83.00	72.68
YCrCb+Gradient	64.55	82.25	72.34
Lab+Gradient	63.68	81.05	71.32
Luv	62.43	84.03	71.64
YCrCb	62.47	82.98	71.28
Lab	61.83	83.98	70.95
RGB	62.09	78.25	69.36

Table 2. Performance (%) comparison of text detection approaches on ICDAR2011 robust reading competition dataset.

Methods	Recall	Precision	Harmonic mean
Our approach	64.64	83.00	72.68
Kims approach [14]	62.47	82.98	71.28
Neumanns approach [10]	64.71	73.10	68.70
Yis approach [13]	58.09	67.22	62.32
TH-TextLoc System [14]	57.68	66.97	61.98
Neumanns approach [8]	52.54	68.93	59.63

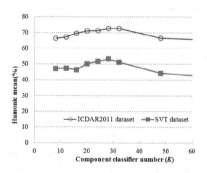

Fig. 6. Performance under different component classifier numbers.

to hundreds and the detection speed increases to 1.6 images per second with performance drop of 4.1 %.

Figure 8 shows some detection examples, where most of the text objects are correctly detected with few false positives. The text objects are in complex background and can have low resolution or low contrast. This shows that the proposed approach can correctly capture text patterns of large variations simultaneously with an integrated discrimination. Figure 8(f) has one miss and Fig. 8(g) two has two missed text regions. The missing text in Fig. 8(f) is due to large distance between the characters. In experiments, it is found that when characters have a

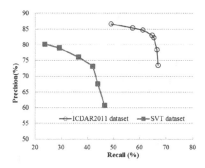

Fig. 7. Curves of recall/precision under different classificiation thresholds.

Table 3. Performance (%) comparison of text detection approaches on SVT dataset.

Color channels	Recall	Precision	Harmonic mean
Our approach	43.89	67.52	53.20
Wangs approach (with lexicon) [16]	29.00	67.00	40.48
Neumanns approach [10]	19.10	32.90	24.17

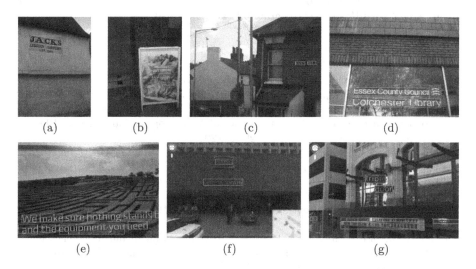

Fig. 8. Detection examples from the ICDAR2011 text dataset (a)-(e), and the Street View Text dataset (f) and (g). Bounding rectanges of text are of green lines and up and down borders of text are indicated by blue lines. (Color figure online)

distance larger than their height, the text objects may be missed. The missing text in Fig. 8(g) is due to the low resolution and perspective deformation.

4   Conclusion and Future Work

Text detection in natural scene images remains a challenging problem due to complex background, low image quality and/or variation of text appearance. In this paper, we develop a discriminative approach that integrates appearance and consensus of components for text detection. We designed a dictionary classifier to discriminate the text components and presented six features that represent the consensus of components. The classifiers are boosted by mining hard negative samples. The text detection is formulated as a hypothesis generation and hypothesis extension process, where the discriminative model is used to control the extension. Experiments have been carried out on two popular datasets to examine the performance of the approach and perform comparisons. Compared with several recent approaches the proposed approach reach the state of the art. Especially, it has a significant performance improvement on the SVT dataset of low quality text objects and clutter backgrounds. Currently, our approach had difficult with multiple touching characters in low resolution images. Vertical text lines or text of deep perspective transformation can also result in missed detection, although the approach works well for most skewed text objects.

Acknowledgement. The partial support of this research by DARPA through BBN/ DARPA Award HR0011-08-C-0004 under subcontract 9500009235, the US Government through NSF Award IIS-0812111 is gratefully acknowledged.

References

1. Liang, J., Doermann, D., Li, H.: Camera-based analysis of text and documents: a survey. Int. J. Doc. Anal. Recogn. **7**, 84–104 (2005)
2. Merino-Gracia, C., Lenc, K., Mirmehdi, M.: A Head-Mounted device for recognizing text in natural scenes. In: Proceedings of Workshop on Camera-Based Document Analysis and Recognition, pp. 29–41 (2011)
3. Yi, C., Tian, Y.: Localizing text in scene images by boundary clustering, stroke segmentation and string fragment classification. IEEE Trans. Image Process. **21**(9), 4256–4268 (2012)
4. Zhao, X., Lin, K.H., Fu, Y., Hu, Y., Liu, Y., Huang, T.S.: Text from corners: a novel approach to detect text and caption in videos. IEEE Trans. Image Process. **20**(3), 790–799 (2011)
5. Phan, T.Q., Shivakumara, P., Tan, C.L.: Text detection in natural scenes using gradient vector flow-guided symmetry. In: Proceedings of the IEEE International Conference Pattern Recognition (2012)
6. Epshtein, B., Ofek, E., Wexler, Y.: Detecting text in natural scenes with stroke width transform. In: Proceedings of the IEEE International Conference, CVPR (2010)
7. Mosleh, A., Bouguila, N., Hamza, A.: Ben: image text detection using a bandlet-Based edge detector and stroke width transform. In: Proceedings of the British Machine Vision Conference (2012)
8. Neumann, L., Matas, J.: Text localization in real-world images using efficiently pruned exhaustive search. In: Proceedings of the International Conference on Document Analysis and Recognition (2011)

9. Chen, H., Tsai, S.S., Schroth, G., Chen, D.M., Grzeszczuk, R., Girod, B.: Robust text detection in natural images with edge-enhanced maximally stable extremal regions. In: Proceedings of the IEEE International Conference on Image Processing (2011)
10. Neumann, L., Matas, J.: Real-time scene text location and recognition. In: Proceedings of the IEEE International Conference on CVPR (2012)
11. Koo, H., Kim, D.H.: Scene text detection via connected component clustering and non-text filtering. IEEE Trans. Image Process. **22**(6), 2296–2305 (2013)
12. Pan, Y., Hou, X., Liu, C.: A hybrid approach to detect and localize texts in natural scene images. IEEE Trans., Image Process. **20**(3), 800–813 (2011)
13. Yi, C., Tian, Y.: Text string detection from natural scenes by structure-based partition and grouping. IEEE Trans. Image Process. **20**(9), 2594–2605 (2011)
14. Shahab, A., Shafait, F., Dengel, A.: ICDAR 2011 robust reading competition challenge 2: reading text in scene images. In: Proceedings of the IEEE International Conference on Document Analysis and Recognition (2011)
15. Lee, J., Lee, P., Lee, S., Yuille, A., Koch, C.: AdaBoost for text detection in natural scene. In: Proceedings of the IEEE International Conference on Document Analysis and Recognition (2011)
16. Wang, K., Babenko, B., Belongie, S.: End-to-End scene text recognition. In: Proceedings of the IEEE International Conference on Computer Vision (2011)
17. Coates, A., Carpenter, B., Case, C., Satheesh, S., Suresh, B., Wang, T., Wu, D.J., Ng Andrew, Y.: Text detection and character recognition in scene images with unsupervised feature learning. In: Proceedings of the IEEE International Conference on Document Analysis and Recognition (2011)
18. Wang, T., Wu, D. J., Coates, A., Andrew, Y.N.: End-to-end text recognition with convolution neural networks. In: Proceedings of the IEEE International Conference on Pattern Recognition (2012)
19. Nister, D., Stewenius, H.: Linear time maximally stable extremal regions. In: Proceedings of the European Conference on Computer Vision (2008)
20. Ye, Q., Han, Z., Jiao, J., Liu, J.: Human detection in images via piecewise linear support vector machines. IEEE Trans. Image Process. **22**(2), 778–789 (2013)

Accuracy Improvement of Viewpoint-Free Scene Character Recognition by Rotation Angle Estimation

Kanta Kuramoto$^{(\boxtimes)}$, Wataru Ohyama$^{(\boxtimes)}$, Tetsushi Wakabayashi, and Fumitaka Kimura

Graduate School of Engineering, Mie University, 1577 Kurimamachiya-cho, Tsu, Mie 514-8507, Japan
{kanta,ohyama,waka,kimura}@hi.info.mie-u.ac.jp
http://www.hi.info.mie-u.ac.jp/

Abstract. This paper addresses the problem of detecting characters in natural scene image. How to correctly discriminate character/non-character is also a very challenging problem. In this paper, we propose new character/non-character discrimination technique using the rotation angle of characters to improve character detection accuracy in natural scene image. In particular, we individually recognize characters and estimate the rotation angle of those characters by our previously reported method and use the rotation angle for character/non-character discrimination. As the result of the character recognition experiment evaluating 50 alphanumeric natural scene images, we have confirmed the accuracy improvement of precision and F-measure by 9.37 % and 4.73 % respectively when compared to the performance with previously reported paper.

Keywords: Rotation-free character recognition · Weighted direction code histogram · Modified quadratic discriminant function · Rotation angle estimation

1 Introduction

Recently, as camera function of cellular phone and digital camera develops, expectation of character recognition by using these as an input device is rising. If such devices capable of recognizing characters in natural scene image were available, they are expected to have a variety of applications. Several applications have been already provided such as Google Goggles and Evernote. Google Goggles can be used for translation system and Evernote can be used for image retrieval. However, these applications still have problem that it cannot detect characters when the characters are not printed along straight line nor rotated individually. Rotation and perspective distortion problem [1] is one of the major challenges on character detection in natural scene image and must be solved in order not to decrease convenience of users.

M. Iwamura and F. Shafait (Eds.): CBDAR 2013, LNCS 8357, pp. 60–70, 2014.
DOI: 10.1007/978-3-319-05167-3_5, © Springer International Publishing Switzerland 2014

There is another problem on character detection in natural scene image. That is, character/non-character discrimination problem [2]. There are many Connected Components (CCs) that are not characters but looks like characters in natural scene images like 'I' and 'L' generated by the edge of object. This problem makes character detection very difficult and must be solved in order to realize high accuracy camera-based OCR.

Many researches have been made in order to deal with rotation and perspective distortion of characters [3–5]. Also, the methods which extract characters from complex scene image by using strong classifier constructed by combining weak classifiers are proposed [6,7]. Moreover, new approach using scene context for character detection in natural scene image is proposed [8]. However not many efforts have been made to deal with character/non-character discrimination problem and any efficient method have not been proposed yet.

In this paper, we propose new character/non-character discrimination technique using the rotation angle of characters to improve character detection accuracy in natural scene image. In particular, we individually recognize characters and estimate the rotation angle of those characters by our previously reported method [9] and use the rotation angle for character/non-character discrimination. There are several rotation angle estimation methods except our previously reported method but most of the methods require at least word unit to estimate its rotation. As the method which can individually estimate character rotation, Uchida et.al. proposed instance-based document skew estimation method [10], but to the best of our knowledge, our method is the first attempt for using rotation angle of characters to improve character detection accuracy in natural scene image.

2   Proposed Method

The processing flow of our method is shown in Figs. 1 and 2. In the learning stage, first we translate the center of the enclosing rectangular of a non-rotated character to the origin of 3D left handed Cartesian coordinate system and the 3D rotation process is performed on a computer to generate 3D rotated characters. After generating 3D rotated characters, feature vectors are extracted from those rotated characters and the mean vector, eigenvalue and eigenvector of covariance matrix are calculated as a learning model for each character class and also for each rotation angle class.

We used Weighted Direction Code Histogram (WDCH) [11] as a feature vector and is extracted by the following procedure.

1. The chain coding is applied to the contour pixels of the normalized character image. Vector sum of adjacent two chain elements is taken to produce 16 directional code.
2. The normalized character image is divided into $(2n-1)^2$ ($(2n-1)$ horizontal $\times$ $(2n-1)$ vertical) blocks. The number of the contour pixels in each direction is counted in each block to produce $(2n-1)^2$ local direction code histograms.

< Learning Stage >

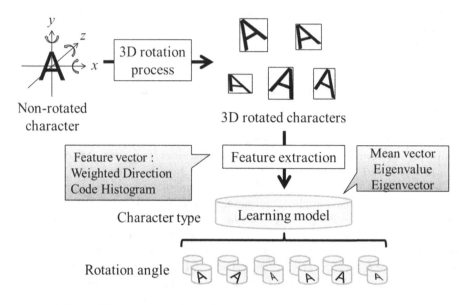

Fig. 1. The processing flow of proposed method (learning stage).

< Recognition Stage >

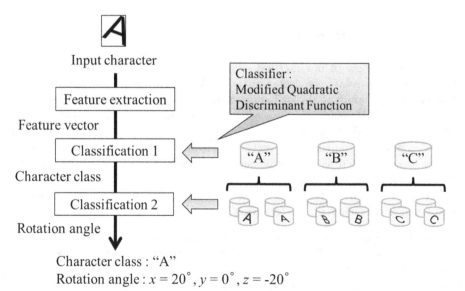

Fig. 2. The processing flow of proposed method (recognition stage).

3. The spatial resolution is reduced from $(2n-1) \times (2n-1)$ to $n \times n$ by down sampling every two horizontal and every two vertical blocks with 5×5 Gaussian filter. Similarly the directional resolution is reduced from 16 to 8 by down sampling with a weight vector $[1\ 2\ 1]^{\mathrm{T}}$, to produce a feature vector of size $8n^2$ (n horizontal, n vertical, and 8 directional resolution).
4. Variable transformation taking square root of each feature element is applied to make the distribution of the features Gaussian-like.

The 5×5 Gaussian filter and the weight vector $[1\ 2\ 1]^{\mathrm{T}}$ in the step 3) are the high-cut filters to reduce the aliasing due to the down sampling. Their size was empirically determined for this purpose.

We used the 392 dimensional WDCH feature vector ($n = 7$) in character recognition and we used the 32 dimensional WDCH feature vector ($n = 2$) in rotation angle estimation.

We classify the 392 dimensional feature vector by using learning model of each character class. As a classifier we used Modified Quadratic Discriminant Function (MQDF) [12].

The MQDF is defined by

$$
g(X) = \frac{1}{\alpha \sigma^2} \left[\|X - M\|^2 \right.
$$

$$
\left. - \sum_{i=1}^{k} \frac{(1-\alpha)\lambda_i}{(1-\alpha)\lambda_i + \alpha\sigma^2} \left\{ \Phi_i^{\mathrm{T}}(X - M) \right\}^2 \right]
$$

$$
+ \sum_{i=1}^{k} \ln\{(1-\alpha)\lambda_i + \alpha\sigma^2\} \qquad (1)
$$

where X and M are feature vector and the mean vector of a class respectively, and λ_i and Φ_i are the ith eigenvalue and eigenvector of the covariance matrix, respectively, $\sigma^2 I$ and α are an initial estimates of the covariance matrix and a confidence constant, respectively. The class which minimizes $g(X)$ is selected as the recognition result. The required computation time and storage is $O(kn)$.

After identifying its character class, we classify the 32 dimensional feature vector by using learning model of each rotation angle class of that character class and identify its rotation angle class. We also used MQDF classifier for this process. Since this approach does not requires rotation normalization of input character, it can recognize rotated characters in the same computational time as non-rotated characters. Also, since this approach recognizes without word/line segmentation characters individual, there is no restriction of text layout. Furthermore, we can identify rotation angle class by using low dimensional feature vector as soon as once character class is identified.

3   Overview

In this section, we overview our character recognition system [9] mainly focusing on the character recognition accuracy.

3.1   Character Recognition System

We introduce our system that automatically detect and recognize the characters in natural scene image. Our system mainly consists of two stages.

The first stage is character segmentation consisting of Connected Component (CC) analysis. The processing flow of this stage is shown in Fig. 3. The input image is binarized by Otsu's method [13], and the CC analysis and noise removal are performed to the binary image. Then enclosing rectangular of the CC is detected and the corresponding area of the original image is extracted. The extracted image is binarized again with the local threshold selected by Otsu's method. We leave only the largest CC to get the most characters likely candidate. We call the CCs detected by the above method as group A (Black characters). As in the similar flow, after the binariztion of the input image by Otsu's method, we perform black and white reverse operation. We call the CCs detected with this reverse operation as group B (White characters). Although more CCs than existing characters are included in the groups, we can extract both black and white characters.

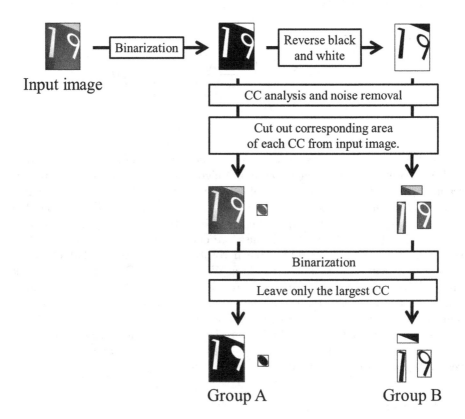

Fig. 3. The processing flow of character segmentation.

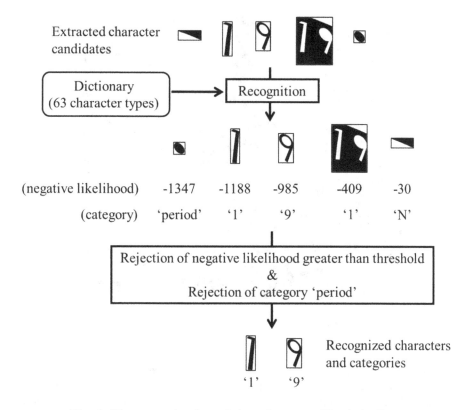

Fig. 4. The processing flow of character recognition/selection.

The next stage is character recognition/selection. The processing flow of this stage is shown in Fig. 4. Since there are CCs that are not characters, we try to reject those CCs that are not character. We use an extended learning model set (with period '.') to calculate the MQDF to recognize characters. The reason of adding period to the category set is to deal with small noise components. We reject CCs that are recognized as period. The value of the MQDF represents the negative likelihood of character and represents that the smaller the value is the higher the possibility of the character. Therefore, if the value is greater than a threshold value, the CC is rejected as it is not a character. We selected threshold value by Otsu's method. Finally, the number of remaining CCs in group A and B are counted up and the components in the group that has more CCs are selected as final characters.

3.2 Character Recognition Accuracy

In order to show the character recognition accuracy of our system, we have done the character recognition experiment using natural scene images for evaluation data. We prepared 50 natural scene images in which there are 1271 printed alphanumeric characters. Some examples of the images are shown in Fig. 5.

Fig. 5. Examples of natural scene images used in the experiment.

We selected printed alphanumeric character as the recognition target excluding disconnected characters 'i' and 'j'. Also, we limited the rotation angle range $-45°$ to $45°$ around x-axis and y-axis, $-30°$ to $30°$ around z-axis in order to avoid similarity problem e.g. 'N' and 'Z'. Therefore, when we generate learning model, we rotated characters within the same rotation angle range discussed above and set the intervals of rotation angle as $15°$. The number 15 is selected by preliminary experiment. We used multi-font data set consisting of approximately 300 samples per class for learning data. Therefore, 73,500 ($300 \times 7 \times 7 \times 5$) samples were generated as learning data for each character class.

As the result of applying our character recognition system to 50 camera-captured images, 4463 CCs that include 1271 characters were extracted in the character segmentation stage. Then, in the character recognition/selection stage, 1223 CCs that should be recognized as a character were correctly recognized, while 199 CCs that should be recognized as not character were detected as a character. Summarized result is shown in Table 1.

Table 1. The results of recognition experiment (number of sample).

Predicted class Actual class	Character	Non-character
Character	1223	48
Non-character	199	2993

Calculating recall (R), precision (P) and F-measure (F) from Table 1, we obtained the results as follows:

$$R = \frac{1223}{1223 + 48} \times 100 = 96.22(\%) \tag{2}$$

$$P = \frac{1223}{1223 + 199} \times 100 = 86.01(\%) \tag{3}$$

$$F = \frac{2 \times recall \times precision}{recall + precision} = 90.83(\%) \tag{4}$$

4 Experiments

4.1 Rotation Angle Estimation Accuracy

In order to show the rotation angle estimation accuracy, we have done the evaluation experiments using artificially generated evaluation data.

In the learning stage, we used the same data set used in Sect. 3 for learning data and prepared three kinds of learning models with rotation intervals of $15°, 10°$ and $5°$ within the same rotation angle range in Sect. 3. The reason we prepared three kinds of learning models is to check how estimation accuracy changes when the intervals of rotation angle gets smaller.

On the other hand in the evaluation stage, we prepared helvetica and century font character for evaluation data. Each character class has one sample and normalized its size to 52×52 pixels by maintaining aspect ratio before feature extraction process. We rotated evaluation data by intervals of $6°$ for x-axis and y-axis within the range of $-42°$ to $42°$, z-axis within the range of $-30°$ to $30°$. Therefore, 2475 ($15 \times 15 \times 11$) samples are generated as an evaluation data set for each character class. When evaluating estimation accuracy, we calculated the arccosine of inner product between the character normal detected by our method and the correct character normal (rotation around x, y-axis). We also calculated the difference between rotation around z-axis to evaluate estimation accuracy.

The results of the experiment is shown in Table 2. Through the experiments, we have found that (1) estimation accuracy improves when the intervals of rotation angle gets smaller, (2) rotation around z-axis is easier to estimate than the rotation around x, y-axis and (3) the font does not affect on the estimation error significantly.

Table 2. The average of estimation error evaluating helvetica and century font character using three kinds of rotation angle learning models ($°$).

	x, y		z	
	Helvetica	Century	Helvetica	Century
15°	14.49	13.73	4.08	3.72
10°	12.84	11.51	3.28	2.70
5°	10.96	9.42	2.18	1.75

The detail estimation error of each character class when using learning models with rotation intervals of 5° is shown in Figs. 6 and 7. We have found that estimation accuracy is lower for characters '0', '1', 'I', 'O', 'f', 'l', 't' and 'y'. This tendency is more significant in Helvetica than in Century.

As common estimation error for both helvetica and century font characters, characters which is slightly rotated around x, y-axis is difficult to estimate its rotation. This is because appearance of character change not much and difficult to tell the difference whether it is rotated or not.

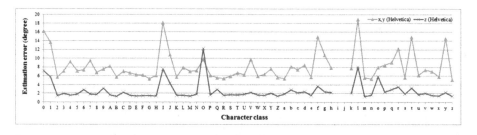

Fig. 6. Estimation error of each character class of Helvetica font.

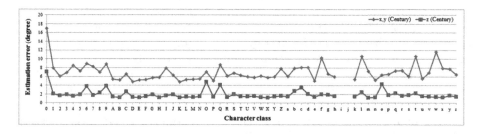

Fig. 7. Estimation error of each character class of Century font.

4.2   Application for Character Detection

From the results of experiment in Sect. IV-A, we have found that our method can precisely estimate the rotation angle of character around z-axis. Therefore, we try to use rotation angle around z-axis to improve character detection accuracy.

We execute the same experiment in Sect. 3. The processing flow of extracting characters from natural scene image is almost the same in Sect. 3 but the only difference is we add CC rejection process using rotation angle around z-axis. We assume that all the characters in the same natural scene image has the same rotation around z-axis. We used learning models with rotation intervals of 10°. The detail process flow of CC rejection using rotation angle around z-axis is shown in Fig. 8. After applying conventional CC rejection process, we count up the number of rotation angle class and detect the class that has highest frequency. We also assume the class within $\pm 10°$ as the correct rotation angle class too. Finally, we leave only the characters that have correct rotation angle class.

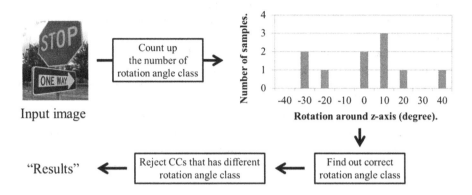

Fig. 8. The process flow of CC rejection using rotation angle.

As the result of the experiment, 1217 CCs that should be recognized as a character were correctly recognized, while 59 CCs that is not character were detected as a character. Summarized result is shown in Table 3.

Table 3. The results of recognition experiment (number of sample).

Predicted class / Actual class	Character	Non-character
Character	1217	54
Non-character	59	3133

Calculating recall (R), precision (P) and F-measure (F) from Table 3, we obtained the results as follows:

$$R = \frac{1217}{1217 + 54} \times 100 = 95.75(\%) \tag{5}$$

$$P = \frac{1217}{1217 + 59} \times 100 = 95.38(\%) \tag{6}$$

$$F = \frac{2 \times recall \times precision}{recall + precision} = 95.56(\%) \tag{7}$$

We obtained 95.75 % recall with 95.38 % precision and the F-measure was 95.56 %. Recall slightly dropped but the precision and F-measure is improved by 9.37 % and 4.73 % respectively. We have confirmed that the rotation angle of characters is one of the useful information to improve character detection accuracy.

5 Conclusion

We evaluated how precisely our previously reported paper can estimate rotation angle of character and also introduced an application of improving character detection accuracy using rotation angle of characters.

As the result of the rotation angle estimation experiments, we have confirmed that our method can estimate rotation angle within $\pm 10°$ around x, y-axis and $\pm 2°$ around z-axis when using learning models with rotation intervals of $5°$. Also, as the result of character recognition experiments using rotation angle of characters, precision and F-measure are improved by 9.37 % and 4.73 % respectively and showed the usefulness of the rotation angle estimation.

While our system can estimate three dimensional rotation, we did not use rotation angle around x-axis and y-axis to improve character detection accuracy. Therefore, future work is to use rotation angle around x-axis and y-axis to improve character detection accuracy.

References

1. Liang, J., Doermann, D., Li, H.: Camera-based analysis of text and documents: a survey. IJDAR **7**(2), 84–104 (2005)
2. Uchida, S.: Challenges in character recognition research. Technical report of IEICE, PRMU2008, vol. 108, no. 432, pp. 49–54 (2009)
3. Myers, G.K., Bolles, R.C., Luong, Q.-T., Herson, J.A., Aradhye, H.B.: Rectification and recognition of text in 3-d scenes. IJDAR **7**(2–3), 147–158 (2005)
4. Iwamura, M., Tsuji, T., Kise, K.: Memory-based recognition of camera-captured characters. In: Proceedings of the DAS2010, pp. 89–96 (2010)
5. Kusachi, Y., Suzuki, A., Ito, N., Arakawa, K.: Kanji recognition in scene images without detection of text fields -robust against variation of viewpoint, contrast and background texture-. In: Proceedings of the ICDAR2001 (2004)
6. Chen, X., Yuille, A.L.: Detecting and reading text in natural scenes. In: Proceedings of the CVPR, vol. 2, pp. 366–373 (2004)
7. Kai-hua, Z., Fei-hu, Q., Ren-jie, J., Li, X.: Automatic character detection and segmentation in natural scene images. J. Zhejiang Univ. Sci. A **8**(1), 63–71 (2007)
8. Kunishige, Y., Feng, Y., Uchida, S.: Character detection from scenery images using scene context. Technical report of IEICE, PRMU2009-221 (2009)
9. Narita, R., Ohyama, W., Wakabayashi, T., Kimura, F.: Three dimensional rotation-free recognition of characters. In: Proceedings of the ICDAR2011, pp. 824–828 (2011)
10. Sakai, S., Uchida, M., Iwamura, M., Omachi, S., Kise, K.: Document skew estimation by instance-based learning. Trans. IEICE **J91–D1**, 136–138 (2008)
11. Kimura, F., Wakabayashi, T., Tsuruoka, S., Miyake, Y.: Improvement of handwritten Japanese character recognition using weighted direction code histogram. Pattern Recogn. **30**(8), 1329–1337 (1997)
12. Kimura, F., Takashina, K., Tsuruoka, S., Miyake, Y.: Modified quadratic discriminant functions and the application to Chinese character recognition. IEEE Trans. Patter. Anal. Mach. Intell. **PAMI–9**(1), 149–153 (1987)
13. Otsu, N.: A threshold selection method from gray-level histograms. IEEE Trans. Syst. Man Cybern. **9**(1), 62–66 (1979)

Sign Detection Based Text Localization in Mobile Device Captured Scene Images

Jing Zhang$^{(\boxtimes)}$ and Rangachar Kasturi

Computer Science and Engineering Department,
University of South Florida, Tampa, FL 33620, USA
jingzhangusf@gmail.com, RlK@cse.usf.edu

Abstract. Sign text is one of the most seen text types appearing in scene images. In this paper, we present a new sign text localization method for scene images captured by mobile device. The candidate characters are first localized by detecting closed boundaries in the image. Then, based on the properties of signboard, the convex regions that contain enough candidate characters are extracted and marked as sign regions. After removing the false positives using the proposed layer analysis, the candidate characters inside the detected sign regions are yielded as sign text. A sign text database with 241 images captured by a mobile device was used to evaluate our method. The experimental results demonstrate the validity of the proposed method.

Keywords: Closed boundary · Text detection · Layer analysis

1 Introduction

With the increasing availability of low cost portable cameras and camcorders, much more pictures and videos are produced than ever before in nowadays. Because these multi-media documents are useful only if they can be navigated efficiently, much effort has been done on Content Based Information Retrieval (CBIR) systems since 1990s. However, the existing CBIR systems are far from perfect [1]. How to index and retrieve the information in images and videos is still a big challenge due to the semantic gap between machine-level and semantic-level descriptors [2]. Fortunately, there is a considerable amount of text occurring in image and video documents. As a well-defined model of concepts for humans' communication, text embedded in multi-media data contains much semantic information related to the content. Therefore, if this text information can be extracted accurately, we can obtain a quite reliable content-based access to the images and videos.

Many text extraction approaches have been proposed since 1990s. Generally, there are four steps of a text extraction system: (1) Text detection, finding regions in an image that contain text; (2) Text localization, grouping text regions into text instances and generating a set of tight bounding boxes around all text objects; (3) Text binarization, binarizing the bounded text and marking text as one binary level and background as the other; (4) Text recognition, performing optical character recognition (OCR) on the binarized text image.

M. Iwamura and F. Shafait (Eds.): CBDAR 2013, LNCS 8357, pp. 71–82, 2014.
DOI: 10.1007/978-3-319-05167-3_6, © Springer International Publishing Switzerland 2014

Sign text is one of the most seen text types appearing in scene images, such as road signs, store signs, bill-boards, and so on. Compared with other text objects, sign text has two important properties: (i) there is a sharp color contrast between sign text and its background in order to let people read it easily; (ii) sign text is located in signboard, which is typically a convex polygon with homogenous color. In this paper, by investigating the properties of both sign text and signboard, we propose a new method that can localize sign text effectively based on sign region detection and layer analysis.

The rest of the paper is organized as follows. Section 2 reviews the related work. Section 3 describes the proposed new sign text localization method. Experimental results are shown and analyzed in Sect. 4. We draw conclusions in Sect. 5.

2 Related Work

Text extraction in image and video documents, as an important research field of content-based information indexing and retrieval, has been developing rapidly since 1990s. Hundreds of approaches have been proposed to address this problem. Chen et al. [3], Jung et al. [4], and Zhang and Kasturi [5] present comprehensive surveys of the text extraction methods. According to the features used and the ways they work, text extraction approaches can be divided into two categories: region based and texture based.

Region based approach utilizes the different region properties between text and background to extract text objects. Color, edge, and connected component are often used in this approach. Shivakumara et al. [6] propose an algorithm to detect video text for low and high contrast images, which are classified by analyzing the edge difference between Sobel and Canny edge detectors. After computing edge and texture features, low-contrast and high-contrast thresholds are used to extract text objects from low and high contrast images separately. Liu et al. [7] use an intensity histogram based filter and an inner distance based shape filter to extract text blocks and eliminate false positives whose intensity histograms are similar to those of their adjoining areas and the components coming from the same object. Bai et al. [8] use a multi-scale Harris-corner based method to extract candidate text blocks. The position similarity and color similarity of Harris corners are used to generate boundaries of text objects. Subramanian et al. [9] use character-stroke to extract text objects. Three line scans defined as a set of pixels along the horizontal line of an intensity image are combined to locate flux regions and flux points, which can show the significant changes of the intensity. Character strokes are detected by using spatial and color-space constraints to find the regions with small variation in color.

Texture based approach uses distinct texture properties of text to extract text objects from background. Machine learning methods are often used in this approach. Hanif et al. [10] extract mean difference feature, standard deviation feature, and histogram of oriented gradient (HOG) feature from the blocks of an image. After AdaBoost algorithm based feature selection, a multilayer perceptron neural network is utilized to detect text objects in scene images. Pan et al. [11] proposed a text detection method by using a conditional random field model. Text regions are detected by HOG feature and cascade boosting classifiers. A conditional random field (CRF) model is used to label candidate text region. Characters are grouped by using minimum spanning tree (MST) and edge cut techniques. Tu et al. [12] calculate the average intensity and statistics of the number of edges from training samples. Adaboost is used to classify the candidate blocks. Text boundaries are matched with pre-generated deformable templates based on shape context and informative features.

Saliency is also studied for text detection task. Shahab et al. [13] evaluate four visual attention-based models and show that attention model can used in early stages of scene text detection. By using saliency map as prior, Uchida et al. [14] use SURF [15] as local features to detect scenery characters. Their experiments demonstrated that SURF and visual saliency can achieve around 75 % accuracy for character and non-character discrimination.

In [16], Clark et al. first use Hough transform and a perceptual grouping method to find the rectangular regions in the scene and then compute edge angle distribution as a confidence measure to find regions that contain text objects. Clark et al. [17] use texture feature and neural network classifier to locate text region. The fronto-parallel view is recovered by using the lines of text paragraph and vanishing point of text plan.

For sign detection and text localization, the method proposed by Bounman et al. [18] first divides the input image into non-overlapping blocks. Then, homogenous blocks detected by luminance thresholding are used as seed to generate homogenous regions by region growing. Finally, holes and contrast are computed for each homogenous region to detect sign regions and holes meeting character criterions in sign regions are output as text. A similar method is presented by Jafri et al. [19]. After obtaining homogenous regions using small non-overlapping blocks and region growing, color distribution is computed to find text regions based on the observation that text regions contain few contrasting colors.

Similar to the methods presented in [16, 17], the proposed sign text detection method also uses the relationship between text object and text plan containing it. However, compared with the dividing and merging strategy-based perceptual grouping methods used in [18, 19], the method presented in this paper can detect sign regions directly based on the color and shape characteristics of sign and, therefore, is robust to sign size and orientation. Furthermore, the proposed layer analysis method can remove false positives effectively by analyzing the properties of text objects.

3   Algorithm

In this section, we introduce the proposed sign text localization algorithm, which is composed of three steps: (i) localize candidate characters by extracting closed boundaries in the image; (ii) find sign regions using color and shape information; (iii) remove false positives outside the detected sign regions and yield the candidate characters using the proposed layer analysis. Figure 1 shows the flowchart of the presented algorithm.

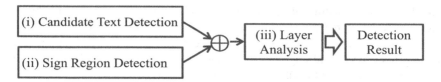

Fig. 1. Flowchart of the algorithm

3.1   Candidate Text Detection

It is observed that, in order to let people read text easily, sign text typically has closed boundary due to the sharp contrast between text and its background. Based on this observation, we are able to localize candidate sign text by detecting closed boundaries in the image.

In our method, the closed boundaries of text are detected by a zero-cross algorithm based edge detector, which can provide closed boundaries more accurately than other detectors in our experiments. Meanwhile, we notice that the edge detector with a low threshold can provide the closed boundaries for sign text but many unexpected weak edges are also detected. Whereas, the edge detector with a high threshold yields strong edges but the boundaries of the sign text may not be closed. In order to detect strong close boundaries and remove noise edges, we generate two edge maps E_l and E_h using a low threshold and a high threshold and combine them as follows:

1. Each boundary in E_l is filtered by its Euler number, which is defined as the number of edge (=1) minus the number of holes of this edge. The closed boundaries are extracted from the boundaries whose Euler numbers are less than 1 (i.e. the boundaries that are have at least one hole).
2. For a extracted closed boundary B in E_l, it is considered as a non-character boundary and removed from E_l, if:

    - (i) The length of B is too short (less than 100 pixels) to be a character boundary;
    - (ii) The aspect ratio (height/width) of the bounding box of B is too large (bigger than 5) or too small (smaller than 0.2);
    - (iii) Less than half of its boundary pixels appear in E_h, which indicates that it is weak boundary.
3. The remaining closed boundaries in E_l form a new edge map E, which contains strong and closed boundaries. These boundaries are marked as candidate text boundaries.

An example is illustrated in Fig. 2. Figure 2-a is an original sign image. Figure 2-b and 2-c are the two edge maps E_h and E_l of Fig. 2-a. We can see that E_h has much less noise edges than E_l, but some boundaries of characters in E_h are not closed. For instance, the character "N" in the word "ONLY" has closed boundary in E_l but unclosed boundary in E_h, as shown in Fig. 2-d and 2-e. Figure 2-f shows the closed boundary map E, which is a combination of the edges in E_l and E_h. Note that although the edge map E contains closed boundaries of all characters, there are still some boundaries of non-character objects in E.

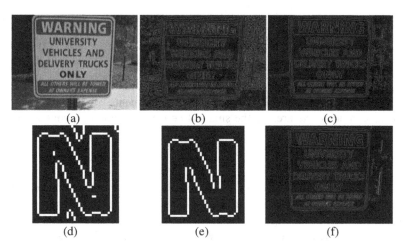

Fig. 2. An example of candidate character detection. (a) original image; (b) weak edge map E_l; (c) strong edge map E_h; (d) a closed boundary in E_l; (e) an unclosed boundary in E_h. (f) edge map E.

3.2 Sign Region Detection

Now the edge map E contains boundaries of all characters and some non-characters. If the sign region can be detected, we are able to extract the character boundaries easily: the boundaries inside the sign region are marked as character boundaries and the boundaries outside the sign region are discarded as non-character boundaries.

As mentioned before, the sign region typically is a convex polygon in homogenous color with several holes (characters). Therefore, we detect sign regions using both color and shape information.

In our method, we use *Statistical Region Merging* (SRM) method proposed by Nock and Nielsen [20] to segment sign regions. SRM uses inference method to model segmentation problem. Statistical test is applied to measure the mean intensities for a small region and its neighbors. They are merged to form one region if their means are similar. The parameter Q in SRM which is used to replace each color channel to Q independent random variables to quantify the statistical complexity was 4 in our experiments.

For a segmented region P generated by SRM, it is considered as a sign region if it satisfies the following three conditions:

1. Let $Num(P_h)$ indicate the number of holes inside P.

$$Num(P_h) > 1 \tag{1}$$

This is because the sign region has homogeneous color and more than one hole due to our assumption thah a text object always contains more than one character.

2. Let A_p indicate the area of P and A_h indicate the total area of holes inside P.

$$A_p + A_h > T_1 \qquad (2)$$

This is because the sign region should be large enough to contain at least two characters. T_1 is set to 100×100 pixels in our experiments.

3. Let A_{mch} indicate the area of the smallest convex polygon that contains P.

$$(A_p + A_h)/A_{mch} > T_2 \qquad (3)$$

This is because we notice that sign regions typically have convex shapes, such as circle, triangle, rectangle, etc. We are able to find the convex regions by computing the area ratio between the sign and the smallest convex polygon containing it. The larger the ratio is, the more the region like a convex polygon. T_2 was set to 0.8 in our experiments.

Figure 3 illustrates the sign region detection of three examples. The first column is the original sign images. The second column shows the segmentation results of SRM. The third column illustrates the detected sign regions using the three conditions described above. We can see that our method can detect sign regions successfully for

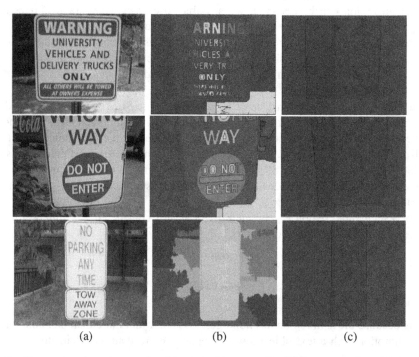

(a)   (b)   (c)

Fig. 3. Sign region detection results. (a) original images; (b) SRM segmentation; (c) detected sign regions.

the images with full sign board, partial sign board, and multiple sign boards. In Sect. 3.3, we will use the detected sign regions to extract sign text.

3.3 Layer Analysis Based Text Detection

In this subsection, we combine the candidate character regions and sign regions detected in the previous subsection to remove false positives and extract characters within the sign region.

After removing the candidate character regions outside the sign region as false positives, the candidate character regions inside the sign regions are labeled using connected component technique and the image is defined as component image C. Figure 4 illustrates the component images of the examples in Fig. 3.

Although all characters are localized, we can see that there are still some non-character components within the sign regions, such as the holes of the characters and two rectangle regions in Fig. 4-a.

In this paper, a new layer analysis method is proposed to remove false alarms and extract characters. A layer image L has the same size as the original image and all pixels in L are initialized to zeros. L is updated using all labeled components iteratively. Assume there are N labeled components within the sign region in C. For the pixels inside the component C_i $i\epsilon[1,...,N]$ (including the holes of C_i), we increase the pixel values at the same locations in L by 1. Consequently, we obtain the final layer image after N iterations. The pseudo code of layer separation can be expressed as below:

```
N: the number of components within the sign region;
Cᵢ: the labeled component in C, iε [1,…,N];
L: the layer image;
Region(Cᵢ): the locations of all pixels inside Cᵢ (includ-
ing the holes of Cᵢ);

L = 0;
FOR  i = 1 to N
        L(Region(Cᵢ)) = L(Region(Cᵢ)) + 1;
END
```

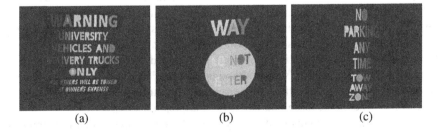

(a) (b) (c)

Fig. 4. Component images of three examples

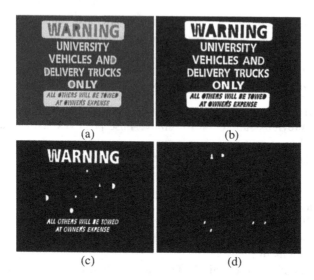

Fig. 5. Layer image and layer separation. (a) layer image; (b) layer 1; (c) layer 2; (d) layer 3

According to the pseudo code, we know that if a pixel is inside m labeled components, the value of the corresponding pixel in L is m. If the maximum value of L is M, we can separate the layer image into M layers using pixel values. The ith layer contains only the pixels whose values are i in L.

The layer image of Fig. 4-a is shown in Fig. 5-a. The components with different layer number are shown in different colors. The regions yellow, red, and brown have layer number 1, 2, and 3, respectively. Figure 5-b, 5-c and 5-d illustrate the separated three layers, among which layer 1 and layer 2 have text objects.

We know that all characters of a sign text object typically are surrounded by the same background. Therefore, the different characters belonging to the same sign text are inside the same number of labeled components. That means the characters of a sign text object always have the same value in L and appear in the same layer.

Based on the obtained layers, we analyze each layer to remove false positive components and find text layers using following heuristics:

- H1: The components with 3 or more holes are removed as non-character components, because English letters have at most 2 holes. For example, the top and bottom components in Fig. 4-b are removed because they have more than 2 holes.
- H2: If a component is surrounded by a component C_i and no other components in this layer are surrounded by C_i, this component is erased as non-character because a sign text contains at least two characters, which are surrounded by the same background. For example, the characters in Fig. 4-b are surrounded by the same component and have the same layer values. While, although the non-characters in Fig. 4-c have the same layer values as well, they are surrounded by different components (different characters). Therefore, these non-character components are erased.
- H3: If a layer has one component or only one left after component elimination using the heuristics H1 and H2, this layer will be discarded as a non-text layer.

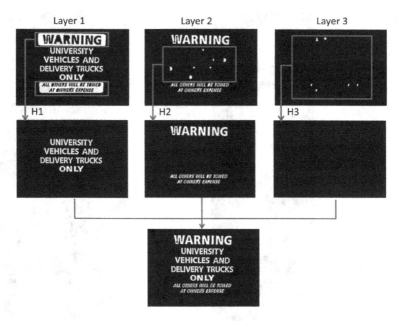

Fig. 6. Layer analysis for text detection

Based on the above heuristics, the non-character components in layer 1 are erased by heuristics H1 and the non-character in layer 2 and layer 3 are erased by heuristics H2, as shown in Fig. 6. We obtain the final text detection result (the bottom image in Fig. 6) by combining the layers after layer analysis.

Finally, the size information and spatial relationship between neighboring characters are used to verify the detected characters, because the characters belonging to the same text object typically have similar sizes and are well aligned along certain direction.

4   Experimental Results

To evaluate the performance of the presented sign text localization method, we tested it using the sign text database released by Bouman et al. [18], which is publicly downloadable at [21]. This database contains total 241 images of road signs, flyers and posters, which were taken by a 0.3-megapixel camera on a Nokia N800 mobile device with resolution 640 × 480. The database is divided into two sets, training set contains 81 images and testing set contains 160 images. The program was implemented by MATLAB in Windows XP operating system.

We selected the parameters and thresholds empirically based on 81 images in training set. For comparison purpose, we computed the standard precision and recall for the 160 images in the testing set. Figure 7 illustrates some results of sign text localization. We can see that our method can detect sign text successfully, including

Fig. 7. Sign text localization results

varying text colors, text sizes, angles of view, orientations, backgrounds, shapes, and partial signs.

The results are listed in Table 1. The proposed sign text detection method outperforms the method presented in [18].

Table 1. Performance evaluation and comparison

Method	Precision	Recall
Our method	*0.837*	*0.878*
Method in [18]	0.818	0.861

Fig. 8. An example that the proposed method failed

Because the proposed method is based on the closed boundary detection, the method may fail when the closed boundaries of character cannot be detected, as the example shown in Fig. 8. Meanwhile, for the signs with occlusions, our method may not provide reliable text regions due to the fact that the occluded sign regions may not have convex shapes.

5 Conclusions

A new sign text localization method has been presented in this paper. Based on our observation that sign text has sharp contrast to the background and sign regions typically have convex shapes with homogenous colors and contain several characters, the proposed method first extracts closed boundaries in the image and label the regions within the boundaries as candidate character components. Then, the convex regions that contain enough candidate character components are detected using both edge and color information and marked as sign regions. After that, the components inside the sign region are filtered by a layer analysis and the remaining candidate components are yielded as sign text.

Our method is evaluated by a sign text database with 241 images using precision and recall measures. The experimental results demonstrate the proposed method can achieve better performance than [18] and the empirical parameters used in the experiments is robust and can detect sign text successfully in most images in the database.

Acknowledgment. The authors wish to thank K.L. Bouman, G. Abdollahian, M. Boutin, and E.J. Delp at Purdue University for proving us the database used in the paper.

References

1. Pavlidis, T.: Limitions of content-based image retrieval. In: Invited Plenary Talk at the 19th International Conference on Pattern Recognition (2008)
2. Hanjalic, A., Lienhart, R., Ma, W.Y., Smith, J.R.: The holy grail of multimedia information retrieval: so close or yet so far away? In: Proceedings of the IEEE, vol. 96, pp. 541–547, April 2008

3. Chen, D., Luettin, J., Shearer, K.: A survey of text detection and recognition in images and videos. Institute Dalle Molled' Intelligence Perceptive (IDIAP) Research Report, August 2000

4. Jung, K., Kim, K.I., Jain, A.K.: Text information extraction in images and video: a survey. Pattern Recogn. **37**(5), 977–997 (2004)

5. Zhang, J., Kasturi, R.: Extraction of text objects in video documents: recent progress. In: Proceedings of International Workshop on Document Analysis and System, pp.5–17 (2008)

6. Shivakumara, P., Huang, W., Phan, T., Tan, C.L.: Accurate video text detection through classification of low and high contrast images. Pattern Recogn. **43**(6), 2165–2185 (2010)

7. Liu, Z., Sarkar, S.: Robust outdoor text detection using text intensity and shape features. In: Proceedings of International Conference on Pattern Recognition (2008)

8. Bai, H., Sun, J., Naoi, S., Katsuyam, Y., Hotta, Y., Fujimoto, K.: Video caption duration extraction. In: Proceedings of International Conference on Pattern Recognition (2008)

9. Subramanian, K., Natajajan, P., Decerbo, M., Castanon, D.: Character-stroke detection for text-localization and extraction. In: Proceedings of International Conference on Document Analysis and Recognition, pp. 33–37 (2007)

10. Hanif, S.M., Prevost, L.: Text detection and localization in complex scene images using constrained adaboost algorithm. In: Proceedings of International Conference on Document Analysis and Recognition, pp.1–5 (2009)

11. Pan, Y., Hou, X., Liu, C.: Text localization in natural scene images based on conditional random field. In: Proceedings of International Conference on Document Analysis and Recognition, pp. 5–10 (2009)

12. Tu, Z., Chen, X., Yuille, A.L., Zhu, S.C.: Image parsing: unifying segmentation, detection, and recognition. Int. J. Comput. Vis. **63**(2), 113–140 (2005)

13. Shahab, A., Shafait, F., Dengel, A., Uchida, S.: How salient is scene text? In: Proceedings of International Workshop on Document Analysis and System (2012)

14. Uchida, S., Shigeyoshi, Y., Kunishige, Y., Feng, Y.: A keypoint-based approach toward scenery character detection. In: Proceedings of International Conference on Document Analysis and Recognition, pp. 819–823 (2011)

15. Bay, H., Tuytelaars, T., Gool, L.: SURF: speeded up robust feature. In: Proceedings of European Conference on Computer Vision (2006)

16. Clark, P., Mirmehdi, M.: Location and recovery of text on oriented surfaces. In: SPIE Conference on Document Recognition and Retrieval, pp. 267–277 (2000)

17. Clark, P., Mirmehdi, M.: Recognising text in real scenes. Int. J. Doc. Anal. Recogn. **4**(4), 243–257 (2002)

18. Bouman, K.L., Abdollahian, G., Boutin, M., Delp, E.J.: A low complexity sign detection and text localization method for mobile applications. IEEE Trans. Multimed. **13**(5), 922–934 (2011)

19. Jafri, S.A.R., Boutin, M., Delp, E.J.: Automatic text area segmentation in natural images. In: Proceedings of International Conference on Image Processing, pp. 3196–3199 (2008)

20. Nock, R., Nielsen, F.: Statistical region merging. IEEE Trans. Pattern Anal. Mach. Intell. **24**, 1452–1458 (2004)

21. http://cobweb.ecn.purdue.edu/~ace/kbsigns/

Font Distribution Observation by Network-Based Analysis

Chihiro Nakamoto[1], Rong Huang[1(✉)], Sota Koizumi[1], Ryosuke Ishida[1], Yaokai Feng[2], and Seiichi Uchida[2]

[1] Graduate School of Information Science and Electrical Engineering, Kyushu University, Fukuoka 819-0395, Japan
{nakamoto,rong,koizumi,ishida}@human.ait.kyushu-u.ac.jp
[2] Faculty of Information Science and Electrical Engineering, Kyushu University, Fukuoka 819-0395, Japan
{fengyk,uchida}@ait.kyushu-u.ac.jp
http://human.ait.kyushu-u.ac.jp/index.html

Abstract. The off-the-shelf Optical Character Recognition (OCR) engines return mediocre performance on the decorative characters which usually appear in natural scenes such as signboards. A reasonable way towards the so-called camera-based OCR is to collect a large-scale font set and analyze the distribution of font samples for realizing some character recognition engine which is tolerant to font shape variations. This paper is concerned with the issue of font distribution analysis by network. Minimum Spanning Tree (MST) is employed to construct font network with respect to Chamfer distance. After clustering, some centrality criterion, namely closeness centrality, eccentricity centrality or betweenness centrality, is introduced for extracting typical font samples. The network structure allows us to observe the font shape transition between any two samples, which is useful to create new fonts and recognize unseen decorative characters. Moreover, unlike the Principal Component Analysis (PCA), the font network fulfills distribution visualization through measuring the dissimilarity between samples rather than the lossy processing of dimensionality reduction. Compared with K-means algorithm, network-based clustering has the ability to preserve small size font clusters which generally consist of samples taking special appearances. Experiments demonstrate that the proposed network-based analysis is an effective way to grasp font distribution, and thus provides helpful information for decorative character recognition.

Keywords: Font distribution · Minimum spanning tree · Centrality criterion · Network-based clustering

1 Introduction

Optical Character Recognition (OCR) techniques have achieved great success in the field of scanner-based document image analysis. However, as demonstrated

M. Iwamura and F. Shafait (Eds.): CBDAR 2013, LNCS 8357, pp. 83–97, 2014.
DOI: 10.1007/978-3-319-05167-3_7, © Springer International Publishing Switzerland 2014

Fig. 1. Scene characters captured by camera.

by Epshtein *et al.* in [1], OCR engines were thwarted in the scene character recognition. This is because the scene character appearing on signboards, notice signage, wrapper, etc. (see Fig. 1), is usually designed by special decoration with the intent to attract people's attention. On the other hand, since camera is far handier than scanner, camera-based OCR, which focuses on recognizing characters captured by camera, will not only extend new applications of OCR but also brings convenience to us in our daily life. In view of this prospect, it has become an imperative demand to develop the camera-based OCR. However, realization of a high-performance camera-based OCR is still a hard task, although numerous impressive methods have been proposed [2–4]. As just mentioned, one of the challenges for scene character recognition lies in the unconstrained appearances with various decorations. Therefore, one possible strategy towards the camera-based OCR is to extract a topological structure that is nearly invariant to decorations.

Along this line of thought, several font-related methods have been elaborately designed as efforts to narrow the gap between OCR and the decorative character. Zhu *et al.* [5] presented a font recognizer by using multichannel Gabor filters and weighted Euclidean distance classifier. Omachi *et al.* [6] detected ridges and ravines from multi-scale images to extract an essential structure of the decorated character. As a subsequent work, Omachi *et al.* [7] matched the graphs of the above-extracted structure and standard patterns to recognize a character image. Unlike relying on the global structure, Wang *et al.* [8] proposed a series of part-based methods which were characterized by the robustness against various appearances of a character.

Although the above methods fulfilled the decorative character recognition to some extents, the performance was far from ideal. A straightforward solution for performance improvement is to collect or enumerate all types of fonts. Unfortunately, this idea is impossible since he/she always can design a new font which takes remarkably different shapes compared with the members of the collected set. As a remedy, we can investigate and analyze a large-scale font set under a certain type of data structure like tree, graph, cluster or network, to grasp the font distribution so that we can approach the ideal effect of brute-force enumeration.

In this paper, we propose a network-based method built on a large-scale font set, which allows us to analyze the font distribution in the feature space. Specifically, the so-called font network is constructed by Minimum Spanning Tree (MST) algorithm taking each font sample as a node. The dissimilarity between two font samples is measured by the Chamfer distance which has been widely adopted in the field like template matching [9] and handwritten Chinese recognition [10]. Merits of our proposal lie in that (1) unlike the well-known Principal Component Analysis (PCA) which lossily projects feature points onto low-dimensional space for distribution visualization, the proposed font network built by linking neighbors can represent the actual font distribution without information loss or distortion; (2) compared with the conventional K-means algorithm, network-based clustering can generate more reliable font cluster and typical samples by introducing some clustering criterion (as introduced in Sect. 3.4). This is because K-means is equivalent to Maximum a Posterior (MAP) estimation of a Gaussian mixture distribution while the font distribution is neither a Gaussian nor a Gaussian mixture; (3) along a path of the font network, we can understand the font shape transition between any two samples, which is useful to create new fonts or recognize various scene characters. For example, for a given decorative character, we can find its neighbours along transition paths of the font network, and then combines multiple recognition results for a final decision. All above mentioned merits are demonstrated by the subsequent experiments.

The remainder of the paper is organized as follows. Section 2 gives an introduction about large-scale font set preparation. Section 3 elaborates the detail procedure of font network construction. In Sect. 4, we conduct experiments and analysis. Section 5 concludes the whole paper and outlines our future works.

2   Large-Scale Font Set Preparation

This section is devoted to a description of large-scale font set preparation. To simplify the problem, the proposed font network is only targeted at the capital alphabet "A" in the current trial. Note that since the process of font network construction is independent of alphabet class, it is feasible and tractable to further accommodate arbitrary alphabet classes. We totally collected 6930 "A"s without font duplication, and normalized each one of them to a 200×200 binary image. Figure 2(a) shows 140 normalized font samples. It should be mentioned that we manually excluded several highly decorative font samples as well as the ones whose main character parts are normal but decorated with various surroundings. We deem this pre-filtering manipulation impartial since even humans may also be hard to make an explicit judgement whether they belong to alphabet or not. See the examples of excluded ones in Fig. 2(b).

3   Font Network Construction

This section introduces the procedure of font network construction based on the prepared large-scale font set. We adopt MST algorithm to build network and

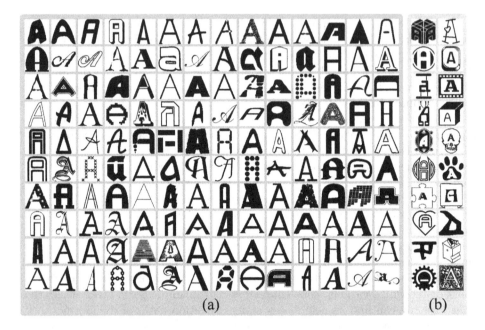

Fig. 2. (a) Examples of normalized font samples. (b) Examples of excluded font samples.

Chamfer distance to measure the dissimilarity between two font samples. To grasp the font distribution, font clusters and typical font samples on the network are then extracted via introducing a distance threshold and some clustering criterion.

3.1   Minimum Spanning Tree

Minimum Spanning Tree (MST) also called minimum weight spanning tree is a pivotal concept in graph theory. Given a connected, undirected graph $G(V, E)$ with vertices $v \in V$ and edges $(v_i, v_j) \in E$ corresponding to pairs of neighboring vertices, a spanning tree $T(V, E')$ of that graph G is a subgraph, namely $E' \subseteq E$, so that all pairs of the vertices are connected by one and only one path. Obviously, a graph can generate many different spanning trees. By assigning a weight $w(v_i, v_j)$ to each edge, MST can then be defined as a spanning tree that has the minimal sum of the weights of the edges E'. In our proposal, font samples in the large-scale set serve as vertices, and the Chamfer distance (see details in Sect. 3.2) between v_i and v_j is regarded as the weight $w(v_i, v_j)$. We adopt Prim's algorithm [11] to construct MST on the large-scale font set, which iteratively adds edges with smallest weights in a greedy matter, and runs in polynomial time.

The advantages of using MST can be summarized as follows: (1) MST not only reflects the global structure of the set via spanning all font samples, but also naturally guarantees that each local edge connects two font samples which are most similar to each other; (2) the path between two vertices allows us to observe the font shape transition; (3) without using dimension reduction projection, MST well preserves the dissimilarity between two font samples and reliably provides

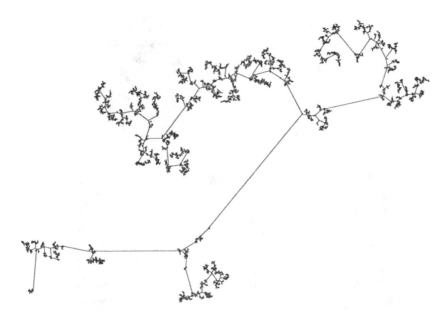

Fig. 3. Visual structure of a font network (MST).

a visual network structure of the font feature space (see Fig. 3); (4) as one type of network structure, MST is compatible with general graph or network theory.

3.2 Chamfer Distance

In this proposal, Chamfer distance [12] is employed to reliably measure the dissimilarity between two font samples and the resulting value serves as the weight $w(v_i, v_j)$ for MST construction. Unlike some naive distance measurement which directly accumulates the absolute difference of pixel intensities, Chamfer distance is average nearest distance from one image to another one so that it is more applicable to shape matching. In the practical algorithm implementation, distance transform is employed to reduce the computational cost of calculating Chamfer distance. This is because distance transform directly stored the wanted nearest distance by labeling each pixel of the image with the distance to the nearest boundary pixel as displayed in Fig. 4. More specifically, given two font samples P and Q which are binary images of size 200×200, p and q denote the corresponding distance maps, respectively. The Chamfer distance is computed as follows:

$$D_{\text{Chamfer}}(P, Q) = \max(d_\alpha, d_\beta)$$

$$d_\alpha = \frac{\sum d(P(i, j); q)}{B(P)} \quad \text{and} \quad d_\beta = \frac{\sum d(Q(i, j); p)}{B(Q)}$$

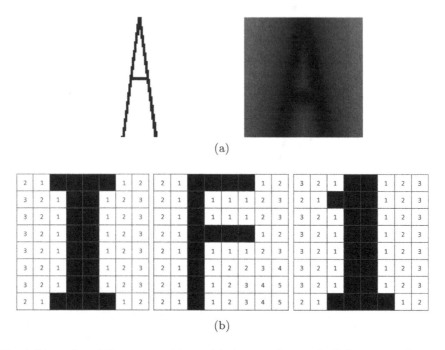

(a)

(b)

Fig. 4. Examples of distance transform. (a) An actual example (left = original image; right = distance map). (b) Artificial examples where the numbers in the table stand for distance values (left = a capital alphabet "I"; middle = a capital alphabet "F"; right = a number "1").

where $B(P)$ counts the number of black pixels of an image P. The operator $d(P(i,j); q)$ is defined as

$$d(P(i,j); q) = \begin{cases} q(i,j) & \text{if } P(i,j) = 0 \text{ (black pixel)} \\ 0 & \text{otherwise} \end{cases}$$

where $P(i,j)$ represents the pixel value in the position (i,j) of an image P. As we can see, the nearest distance can be obtained by visiting the same position of the corresponding distance map. Thus, distance transform helps reduce the computational complexity to the level of look-up table in the course of calculation.

3.3   Network-Based Clustering

To well grasp the font distribution, we propose a network-based clustering, by which font samples with short Chamfer distances to each other are grouped together. In doing so, a coarser overview of the original font network (MST) can be obtained that allows us to investigate its global configuration on different scales. The local details are reflected in the fact that font samples within a cluster share similar shape. Note that both the global configuration and the local details indicate the distribution of the large-scale font set. Moreover, to

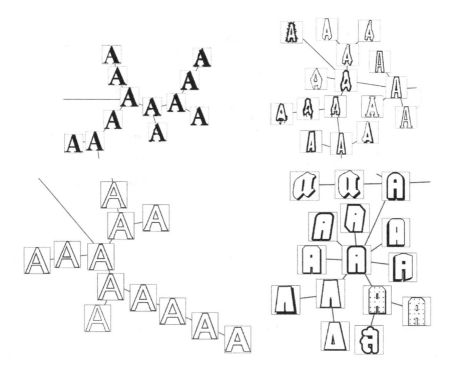

Fig. 5. The internal structures of four clusters.

effectively represent and observe font clusters, we utilize some centrality criterion (see details in Sect. 3.4) to extract typical font samples.

The clustering starts with setting a distance threshold T_D. Then, traverse the vertices throughout the font network. If $w(v_i, v_j) \leq T_D$, the vertices v_i and v_j are grouped into a same font cluster.

Network-based clustering preserves the internal structure of each cluster as shown in Fig. 5, and thus allows us to locally observe the font shape transition even after the clustering. More importantly, small size clusters survive from the network-based clustering as long as these clusters are essentially away from others in terms of Chamfer distance. On the contrary, K-means algorithm is approximately equally divided the whole feature space into several clusters so that scattered font samples with special shape are forced to merge into dissimilar large clusters. This contrast is illustrated in Fig. 6 and a subsequent experiment. Here, the size of a cluster refers to the number of samples belonging to that cluster.

3.4   Typical Font Sample Extraction by Centrality Criterion

To extract a typical font sample from each cluster, this subsection introduces three centrality criteria, namely closeness centrality, eccentricity centrality and betweenness centrality, which are widely employed in the network analysis [13]. The centrality criterion estimates the degree of center for each vertex, and returns

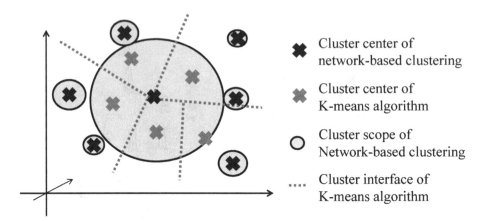

Fig. 6. An illustration indicating the different performance between the network-based clustering and the K-means algorithm. Experimental results of comparative study can be found in Table 1.

comparable scores. In this proposal, we adopt the centrality criterion to extract the typical font sample from each cluster. In the following, three centrality criteria are explained in turn.

Closeness centrality is based on the natural distance metric between all pairs of vertices as given below.

$$C_C(v_i) = \frac{n-1}{\sum_{j=1}^n d(v_i, v_j)}, \quad i = 1, 2, \cdots, n,$$

where n denotes the size of a font cluster. The distance $d(v_i, v_j)$ takes the sum of $w(v_i, v_j)$ along the path from v_i to v_j. Recall that $w(v_i, v_j)$ is the Chamfer distance between two directly connected vertices. A vertex v_i with largest $C_C(v_i)$ is considered as the typical font sample.

Eccentricity centrality selects the center sample by comparing all pairs of maximum distances. That is

$$C_E(v_i) = \frac{1}{\max d(v_i, v_j)}, \quad i, j = 1, 2, \cdots, n.$$

The denominator $\max d(v_i, v_j)$ can be defined as the degree of eccentricity so that a larger $C_E(v_i)$ indicates a more compact extent that the vertices v_j (where $j = 1, 2, \cdots, n$) gather around v_i.

Betweenness centrality quantifies the number of times that a vertex acts as a bridge along the path between two other vertices. More specifically, the betweenness centrality can be represented as follows.

$$C_B(v_i) = \sum_{s \neq i \neq t = 1}^n \frac{\sigma_{v_s v_t}(v_i)}{\sigma_{v_s v_t}}, \quad i = 1, 2, \cdots, n,$$

where $\sigma_{v_s v_t}$ is total number of edges from v_s to v_t ($\forall s \neq t \in \{1, 2, \cdots, n\}$) and $\sigma_{v_s v_t}(v_i)$ accumulates the number of times that all these paths pass through v_i.

The betweenness centrality relies on the natural fact that the center vertex has a greater opportunity to be passed through by paths. Therefore, the vertex v_i having the largest $C_B(v_i)$ corresponds to the typical font sample.

4 Experiment and Analysis

4.1 Font Network

In this experiment section, the MST algorithm and the Chamfer distance were applied to construct the font network of the large-scale alphabet "A" set. The global structure of the built network was displayed in Fig. 3. Further, according to the metric of dissimilarity, the network-based clustering algorithm divided the feature space into several clusters without affecting their internal structures (see Fig. 5). In addition, Fig. 7 exhibited the font shape transition between two vertices, which was useful to generate new fonts or recognize various scene characters. It was worthwhile to point out that we could still observe the font shape transition even after the clustering processing, which benefited from the above mentioned structure preservation property. Moreover, after introducing some centrality criterion like closeness centrality, eccentricity centrality or betweenness centrality, the typical font sample could be extracted from each cluster. Note that the typical font sample provided an effective representation for each

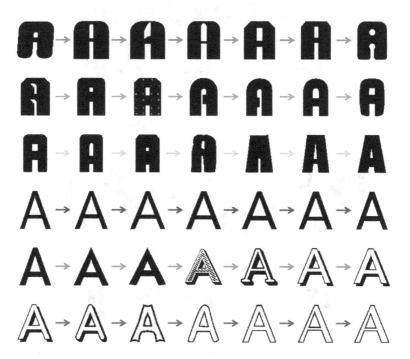

Fig. 7. Font shape transition along paths.

font cluster. All above investigations allowed us to grasp the distribution of a large number of font samples.

4.2 Vertex Degree Histogram

The degree of a vertex is defined as the number of edges incident to the vertex. Figure 8 showed the histogram of vertex degree and provided another aspect of the font distribution. As we could see, the maximum vertex degree is eight. In addition, the number of vertices diminished with the increase of degree, which

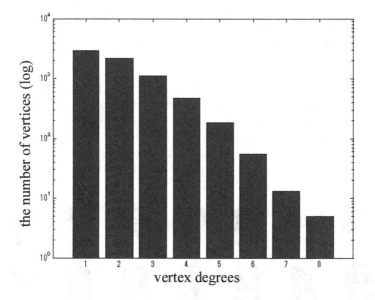

Fig. 8. Vertex degree histogram.

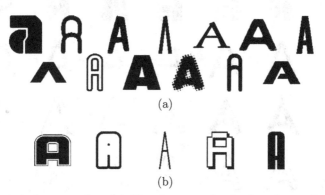

Fig. 9. The hub font samples. (a) The font samples having degree 7; (b) the font samples having degree 8.

implied that the vertices having highest degree could be considered as the hubs of a network. See the hub font samples in Fig. 9. Note that the hub font samples were different from the typical ones. The former was selected from the whole network and reflected the global distribution while the latter was the local representative extracted from each font cluster using some centrality criterion.

4.3   Font Cluster

Since the distance threshold T_D had a great impact on clustering, in this subsection, we analyzed the configuration evolution of font clusters. We displayed the evolution of the maximum size and the number of clusters versus the increase of T_D in Fig. 10, which demonstrated the existence of the font cluster.

Table 1 listed the size of the top five largest font clusters with respect to each distance threshold T_D. Furthermore, the result of K-means algorithm ($K = 5$) was given in the last row of Table 1. The results listed in Table 1, also illustrated in Fig. 6, indicated that the network-based clustering could preserve small size font clusters which contained samples taking special shapes, while K-means approximatively equally divided the feature space.

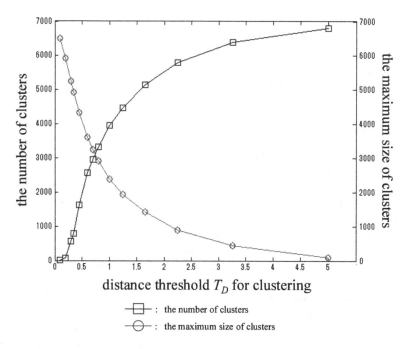

Fig. 10. The evolution of the maximum size and the number of clusters versus the increase of T_D.

Table 1. The size of font clusters.

	T_D	1st	2nd	3rd	4th	5th
	0.10	10	8	7	6	5
	0.20	67	66	37	32	23
	0.30	566	114	33	26	24
	0.35	796	238	63	21	18
	0.45	1630	26	22	22	22
	0.60	2560	19	18	17	16
Network-Based Clustering	0.70	2953	25	18	18	17
	0.80	3330	42	18	17	17
	1.00	3957	79	17	14	10
	1.25	4469	97	17	10	10
	1.65	5149	23	10	10	10
	2.25	5816	28	11	8	8
	3.25	6403	6	6	5	5
	5.00	6813	5	3	3	2
K-means	$K = 5$	2128	1444	1308	1128	922

4.4   Typical Font Sample

Typical font samples were extracted by introducing some centrality criterion, namely closeness centrality, eccentricity centrality or betweenness centrality, as described in Sect. 3.4. The selected font samples from the top five largest clusters were exhibited in Fig. 11. The digits printed above each sub-picture stood for the size of the cluster from which the typical font sample was extracted. As we could observe, when the distance threshold T_D was small, for example $T_D = 0.30$, the typical font samples shared small dissimilarity to each other. With the increase of T_D, the great dissimilarity among the typical font samples emerged as shown in Fig. 11 (c). Moreover, the typical font sample from the largest cluster shared the similar standard shape regardless the change of T_D, which indicated that no matter how protean a font sample was, its appearance would hold approximately constant structures. In addition, applying three centrality criteria leaded to similar results. In other words, centrality criterion would not sharply affect the process of typical font sample extraction.

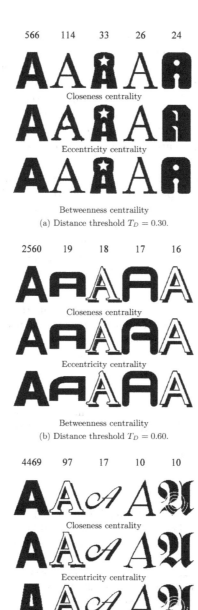

(a) Distance threshold $T_D = 0.30$.

(b) Distance threshold $T_D = 0.60$.

(c) Distance threshold $T_D = 1.25$.

Fig. 11. The typical font samples extracted by centrality criterion.

5   Conclusion

In this paper, we analyze the font distribution of a large-scale set by network, which opens a new door to the camera-based OCR engines. To construct the font network, we adopt MST algorithm under the dissimilarity measurement using Chamfer distance. Font clusters are formed though setting distance threshold. After that, we extract typical font samples from clusters by introducing some centrality criterion, namely closeness centrality, eccentricity centrality and betweenness centrality. Benefitting from the network structure, both the global configuration and the font shape transition can be observed. Compared with the conventional PCA, the proposed font network realizes distribution visualization through Chamfer distance rather than the process of dimensionality reduction. Moreover, as verified by experiments, the network-based clustering preserves small size font clusters, while K-means algorithm will produce an approximately equal division. The existence of font cluster and the effectiveness of network-based analysis are also demonstrated by experiments. Our future work is to extract the internal structures from font clusters, and to design regularization approaches based on the path of font shape transition.

Acknowledgment. The authors would like to thank the support of Creation of Human-Harmonized Information Technology for Convivial Society, which is a CREST project organized by Japan Science and Technology Agency (JST).

References

1. Epshtein, B., Ofek, E., Wexler, Y.: Detecting text in natural scenes with stroke width transform. In: IEEE Conference on Computer Vision and Pattern Recognition (2010)
2. Wang, K., Belongie, S.: Word spotting in the wild. In: Daniilidis, Kostas, Maragos, Petros, Paragios, Nikos (eds.) ECCV 2010, Part I. LNCS, vol. 6311, pp. 591–604. Springer, Heidelberg (2010)
3. Wang, K., Babenko, B., Belongie, S.: End-to-end scene text recognition. In: IEEE International Conference on Computer Vision (2011)
4. Mishra, A., Alahari, K., Jawahar, C.: Top-down and bottom-up cues for scene text recognition. In: IEEE Conference on Computer Vision and Pattern Recognition (2012)
5. Zhu, Y., Tan, T., Wang, Y.: Font recognition based on global texture analysis. IEEE Trans. Pattern Anal. Mach. Intell. **23**(10), 1192–1200 (2001)
6. Omachi, S., Inoue, M., Aso, H.: Structure extraction from decorated characters using multiscale images. IEEE Trans. Pattern Anal. Mach. Intell. **23**(3), 315–322 (2001)
7. Omachi, S., Megawa, S., Aso, H.: Decorative character recognition by graph matching. IEICE Trans. Inf. Syst. **E90–D**(10), 1720–1723 (2007)
8. Wang, S., Uchida, S., Liwicki, M.: Part-based recognition of arbitrary fonts. In: International Conference on Document Analysis and Recognition (2013)
9. Borgefors, G.: Hierarchical chamfer matching: a parametric edge matching algorithm. IEEE Trans. Pattern Anal. Mach. Intell. **10**(6), 849–865 (1988)

10. Shi, D., Gumm, S., Damper, R.: Handwritten chinese radical recognition using nonlinear active shape models. IEEE Trans. Pattern Anal. Mach. Intell. **25**(2), 277–280 (2003)
11. Cormen, T., Leiserson, C., Rivest, R., Stein, C.: Graph algorithms (Section 23.2: The algorithms of Kruskal and Prim). In: Cormen, T.H. (ed.) Introudction to Algorithms, 3rd edn, pp. 631–638. MIT Press, Cambridge (2009)
12. Barrow, H., Tenenbaum, J., Bolles, R., Wolf, H.: Parametric correspondence and chamfer matching: two new techniques for image matching. In: International Joint Conference Artificial Intelligence (1977)
13. Opsahl, T., Agneessens, F., Skvoretz, J.: Node centrality in weighted networks: generalizing degree and shortest paths. Soc. Netw. **32**(3), 245–251 (2010)

Camera-Based Systems

Dewarping Book Page Spreads Captured with a Mobile Phone Camera

Chelhwon Kim[1]([✉]), Patrick Chiu[2], and Surendar Chandra[2]

[1] Electrical Engineering Department, University of California, Santa Cruz, CA, USA
chkim@soe.ucsc.edu
[2] FX Palo Alto Laboratory, Palo Alto, CA, USA
{chiu,chandra}@fxpal.com

Abstract. Capturing book images is more convenient with a mobile phone camera than with more specialized flat-bed scanners or 3D capture devices. We built an application for the iPhone 4S that captures a sequence of hi-res (8 MP) images of a page spread as the user sweeps the device across the book. To do the 3D dewarping, we implemented two algorithms: optical flow (OF) and structure from motion (SfM). Making further use of the image sequence, we examined the potential of multi-frame OCR. Preliminary evaluation on a small set of data shows that OF and SfM had comparable OCR performance for both single-frame and multi-frame techniques, and that multi-frame was substantially better than single-frame. The computation time was much less for OF than for SfM.

Keywords: Document capture · Document analysis · Dewarping · Mobile phone camera · Book scanning

1 Introduction

Using portable devices to capture images of documents is a fast and convenient way to scan documents. Being able to use the compact capture device on-site is an important benefit in many scenarios. For example, students can use them to copy pages from books in a library, without potentially damaging the book spines when copying with a flat-bed copier. Another example is the digitization of documents in storage, in which bounded or loose paper records are often in too poor a condition to be used with flat-bed or V-bed book scanners without damaging them.

Compared with the results produced by flatbed scanners, these photos of documents taken with portable devices suffer from various issues including perspective distortion, warping, uneven lighting, etc. These defects are visually unpleasant and are impediments to OCR (optical character recognition). This paper focuses on the problem of dewarping page spread images of a book captured by a hi-res mobile phone camera.

We built an app for the iPhone 4S, which has an excellent camera, to capture a sequence of frames (8 MP, 2 fps). To capture a page spread, the user

M. Iwamura and F. Shafait (Eds.): CBDAR 2013, LNCS 8357, pp. 101–112, 2014.
DOI: 10.1007/978-3-319-05167-3_8, © Springer International Publishing Switzerland 2014

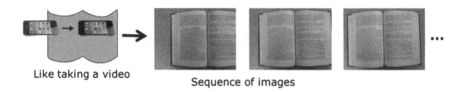

Like taking a video

Sequence of images

Fig. 1. Capturing a page spread of a book.

simply sweeps the device across the open book, similar to taking a video (see Fig. 1). From the sequence of frame images, we estimate the 3D information. We have implemented both optical flow (OF) and structure from motion (SfM) algorithms. The output of this step is a disparity map which encodes the depth information. Then we leverage the dewarping module in our previous system (where the disparity map was obtained from a stereo camera) [7]. This dewarping algorithm uses a 3D cylindrical model. An overview of the pipeline is illustrated in Fig. 2.

Making further use of the sequence of frame images, we consider a multi-frame OCR approach to improve the OCR performance. The idea is based on the observation that the left and right pages may be in better focus and not cropped off in different frames as the phone camera sweeps across the page spread at a non-uniform velocity.

We performed a preliminary evaluation to compare the OF and SfM algorithms in terms of OCR performance and computation time. We also compared multi-frame OCR with single-frame OCR using the middle frame image to see whether the improvement is substantial. The results are reported in detail below.

2   Related Work

Existing research systems have been developed that relies on special 3D cameras or mounting hardware. The Decapod system [15] uses two regular cameras with special mounting hardware. Our previous system [7] uses a consumer-grade compact 3D stereo camera (Fujifilm Finepix W3). The dewarping method in our system is based on a cylindrical model, which for non-3D images performed the best (though the difference was not statistically significant) in the Document Image Dewarping Contest at CBDAR 2007 (see [8,14]).

Other 3D capture devices include structured light, which can sense highly accurate 3D information but requires more complicated apparatus. An example system is [4].

While it is possible to dewarp a book page image from a single photo taken with a non-3D device, the techniques to compute the 3D information are more specialized. Approaches include detecting content features like curved text lines or page boundaries and then applying a 3D geometric model to dewarp the image (e.g. [5,6,8,9]).

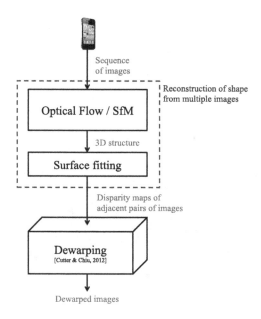

Fig. 2. Pipeline of system.

Using video to capture documents is perhaps the approach that is the most related to our present work. With standard video formats, the frame image resolution is limited (VGA at 0.3 MP, HD at 2 MP) and performing OCR is problematic. In contrast, our app captures frames at much higher resolution (8 MP).

An early system, Xerox XRCE CamWorks ([11,18]), has a video camera mounted over a desk to capture text segments from flat documents. It applied super-resolution techniques and OCR was evaluated on simulated images but not on actual camera images.

The NEC system [10] uses a VGA webcam and a mobile PC to capture video of a flat document or a curved book page spread. The user sweeps over the document in a back-and-forth path in order to cover the document and an image mosaicing method is applied to reconstruct an image of the whole document. The mosaicing uses a structure from motion algorithm that tracks Harris corner feature points. OCR was not performed nor evaluated.

Our system also uses a structure from motion algorithm that tracks Good Features To Track (GFTT) feature points [16]. In addition, we implemented a simpler optical flow algorithm. The high resolution allows us to use optical flow because a single sweep can capture the whole image and mosaicing is not need. Mosaicing requires a global coordinate system that SfM computes but OF does not. With OF, it suffices that only adjacent pairs of frames share a consistent coordinate system.

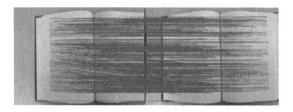

Fig. 3. Identifying corresponding feature points.

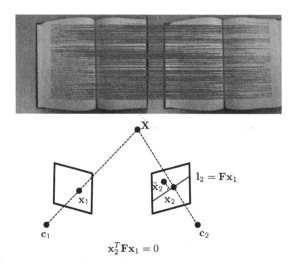

$$l_2 = Fx_1$$

$$x_2^T F x_1 = 0$$

Fig. 4. Removing outliers using epipolar geometry.

3   Computing and Dewarping the 3D Structure

We proceed to describe our implementation of two methods to compute the 3D structure: optical flow (OF) and structure from motion (SfM). In both, the features that are tracked are GFTT feature points [16]. Another option for feature points is the popular SIFT points; however SIFT points are not specifically designed to be tracked like the GFTT points. We also perform camera calibration to model the cameras geometry and correct for the lens distortions, which depends on the individual iPhone 4S device. The algorithms for GFTT and camera calibration are available in the OpenCV [3] computer vision library. The output of these OF and SfM methods is a disparity map that encodes the depth information, which are then fed into the dewarping module in the pipeline (see Fig. 2).

3.1   Optical Flow

First, for each pair of sequential frame images, the corresponding feature points are matched. An example is shown in Fig. 3.

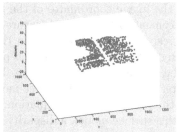

(a) Optical flow disparities (upper-left corner shows a closeup).

(b) Recovering shape information.

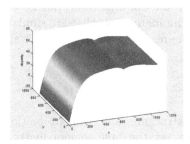

(c) Surface fitting.

(d) Disparity map with document region localized.

Fig. 5. Computing disparity map from optical flow.

Next, the outliers are removed using epipolar geometry between two frames, which is described in the following equation

$$\mathbf{x}_2^T \mathbf{F} \mathbf{x}_1 = 0 \qquad (1)$$

where $\mathbf{F}$ is the fundamental matrix, $\mathbf{x}_1$ and $\mathbf{x}_2$ are homogeneous coordinates of the projected points of 3D point $\mathbf{X}$ onto the first and second image plane respectively. From this equation, we can map $\mathbf{x}_1$ to a line $\mathbf{l}_2 = \mathbf{F}\mathbf{x}_1$ in the second image. In other words, the projected point $\mathbf{x}_2$ on the second image plane always lies on the line. However, we cannot guarantee that all pairs of corresponding feature points satisfy this epipolar constraint due to noise in the image measurements and error in the optical flow matching method.

Therefore, to identify outliers among them, we calculate the orthogonal distance from the matching point in the second image, $\tilde{\mathbf{x}}_2$ to $\mathbf{l}_2$ (see Fig. 4), and if the distance is beyond a certain threshold then the pair of corresponding points is considered as an outlier. Figure 4 shows the remaining inliers.

Computing disparities from optical flow is accomplished by looking at the displacements of the tracked feature points. The points on the book page spread at different depths will have different displacements (Fig. 5(a)), and these disparities can be used to recover the shape of the page spread (see Fig. 5(b)). Each dot in Fig. 5(b) represents a pair of corresponding points in the 3D space, where (x, y) are the image coordinates of the feature point in the first image, and z is the displacement of the tracked feature point in the second image with respect to the corresponding feature point in the first image. The recovered 3D points are clustered into two groups on each page; currently this is done manually by labeling the location of the book spine. This process can be automated by applying a clustering algorithm. A surface model is fitted to each cluster of 3D points using a 4-th order polynomial equation. See Fig. 5(c).

From this surface model, a disparity map is generated by mapping the depth (z-coordinate) to a grayscale value. Finally, the document region is localized within the image using an image segmentation algorithm; a good algorithm is GrabCut [13], which is available in OpenCV. In order to apply GrabCut, some background pixels must be identified and one way to do this is to sample pixels around the edge of the image and eliminate those that are similar to the center area of the image. An example of the resulting disparity map is shown in Fig. 5(d).

3.2   Structure Form Motion

The first step is to initialize the 3D structure and camera motion from two sequential frames as follows: we first set the first camera matrix $\mathbf{P}_1 = \mathbf{K}[\mathbf{I}_{3\times3}|\mathbf{0}_{3\times1}]$ to be aligned with the world coordinate frame, where $\mathbf{K}$ is the camera calibration matrix. Next, we identify the corresponding points between those two frames and estimate the fundamental matrix $\mathbf{F}$ using RANSAC algorithm. This is available in OpenCV library. The fundamental matrix is used to remove outliers as described above. Then, the essential matrix is computed by $\mathbf{E} = \mathbf{K}^T\mathbf{F}\mathbf{K}$. Once we have determined the essential matrix, we can recover the camera pose (rotation $\mathbf{R}$ and translation $\mathbf{t}$) for the second frame with respect to the first camera frame [17]. Then $\mathbf{P}_2$, the camera matrix for the second frame, can be easily obtained by multiplying the camera calibration matrix $\mathbf{K}$ by the camera pose for the second frame $[\mathbf{R}|\mathbf{t}]$. Lastly, we estimate the 3D point structure from the 2D corresponding points and $\mathbf{P}_2$ through triangulation [17].

In practice, the algorithm for the fundamental matrix might not produce a well-conditioned initial 3D structure due to noise in the image measurements. Therefore, we add a step to reject ill-conditioned structures. An example of an ill-conditioned initial 3D structure is shown in Fig. 6(a). The criterion of rejection is based on the prior knowledge that the shape of a book spread page is almost always two slightly curved surfaces that are not too far from a plane. Therefore, we first detect a dominant plane using RANSAC from the generated 3D structure, and then calculate the orthogonal distance for each 3D point to the plane. If the average distance is less than a predefined threshold then we accept the pair of frames, or reject it and check the next pair of frames. The threshold can be fixed under an assumption that the distance between the camera and the target is almost consistent across different users. Figure 6(b) shows a well-conditioned 3D structure from the selected pair of frames.

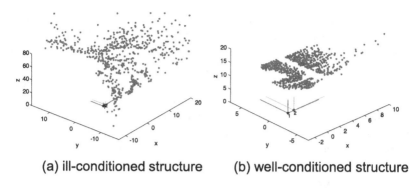

(a) ill-conditioned structure   (b) well-conditioned structure

Fig. 6. Initial 3D structure.

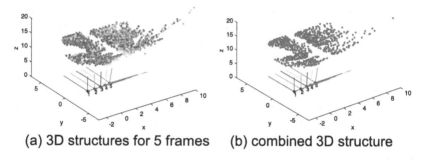

(a) 3D structures for 5 frames   (b) combined 3D structure

Fig. 7. Structure from motion: after 5 frames.

An alternative method for computing the fundamental matrix is to use a non-linear optimization technique (e.g. [1]). This might improve the accuracy of the camera pose, but it requires more complicated processing.

Now we have an initial 3D point structure and consider how to use a new frame to update it. Let us assume that the 3D point structure for $(i-1)$-th frame is already known and we have tracked the existing corresponding points from the $(i-1)$-th frame to the ith frame. As we described above, we remove outliers from the tracked points using epipolar geometry. The remaining tracked points and the corresponding 3D points are used to estimate the new camera pose for ith image $\mathbf{P}_i$ by minimizing the projection error $e = \sum_j \left\| \mathbf{x}_j^{(i)} - \mathbf{P}_i \mathbf{X}_j \right\|^2$, where $\mathbf{x}_j^{(i)}$ is the jth tracked 2D point in the ith image and $\mathbf{X}_j$ is the corresponding jth 3D point. Given this estimated camera matrix $\mathbf{P}_i$ and the tracked points in the ith frame, we recalculate the 3D point structure through triangulation. We iterate the above process throughout the sequence of frames. Figure 7(a) shows the 3D point structures for each iteration and camera pose frames with different colors. To get a single 3D structure from all the frames 3D structures, we combined them by simple averaging (Fig. 7(b)). The final 3D structure still has outliers as can be seen from the right most corner of the structure in Fig. 7(b). In order to deal with this, we perform the surface fitting algorithm with RANSAC.

From the surface model, a disparity map is generated for each frame as described above in the optical flow method.

Another option for combining all the 3D structures is to use bundle adjustment (e.g. [20]). The advantage is that it might improve the accuracy of the camera poses and the 3D structures. Since in our application, the camera motion is very simple (basically linear), the improvement may be small. The disadvantage of using bundle adjustment is that it requires more processing.

3.3 Cylindrical Model

For completeness, we give a brief summary of how the cylindrical model is used with the disparity map to do the dewarping; for more details refer to [7]. First, from a disparity map, two depth profiles perpendicular to the spine are extracted from the top and bottom halves of the page spread by averaging over their respective halves. These profiles form the skeleton of the cylindrical model. To facilitate the rendering of the dewarped image, rectangular meshes are employed. A mesh vertex point on the cylindrical model can be mapped to a vertex point in the dewarped image by flattening it using its arclength along the cylindrical surface to push it down and outward from the spine. Points inside each mesh rectangle are then interpolated based on the rectangle's vertices.

4   Multi-frame OCR

By *single-frame* OCR, we mean using one frame to OCR the left and right pages of a page spread. Typically, the middle frame in the sequence of frame images can be used, because both pages of the book spread are usually in view with the camera held in landscape orientation.

By *multi-frame* OCR, we mean using more than one frame for doing the OCR. The idea is that the left page is more likely to be better captured in the early frames and the right page in the later frames. Some frames may also be in better focus than others.

To study the potential of multi-frame OCR, we compared the best OCR scores for the left and right pages over multiple frames to the OCR scores of the middle frame. These results are reported below.

For single-frame OCR and multi-frame OCR, a separate condition is whether the frame images have been dewarped.

5   Preliminary Evaluation

To compare OF vs. SfM, non-dewarped vs. dewarped, and single-frame vs. multi-frame, we did a preliminary evaluation on a small set of data based on OCR.

Six images of book page spreads were taken with our app on an iPhone 4S camera. The device was handheld (a tripod was not used). The frame image

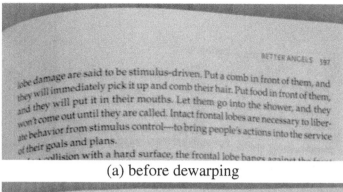

(a) before dewarping

(b) after dewarping

Fig. 8. Example of a dewarped page spread with the top-right region shown.

resolution was 8 MP (3264×2448). The frame rate used was about 1 fps; we found this to work fine for our processing pipeline even though the frame rate can go as high as 2 fps when capturing 8 MP images.

Our mobile phone app was implemented in Objective-C, and the code for processing the frame images was implemented in C++ and uses the OpenCV library [3]. The captured images were processed on a desktop PC.

We examined the boundary text lines on the two pages in each page spread: top-left, top-right, bottom-left, bottom-right. By a boundary text line, we mean the text line nearest to the top or bottom of a page that spans more than half the body of the page, so that short lines at the bottom of a paragraph and headers or footers are not considered. The 6 page spreads provides a total of 24 boundary text-lines.

An example of a dewarped page spread is shown in Fig. 8. The frame is the middle frame of the image sequence. The method is OF. The bottom image in the figure is a closeup of the top-right region of the page spread showing several text lines that have been dewarped. There is some inaccuracy near the spine of the book, which is a difficult area to handle due to the steepness of the page and the lack of content for tracking.

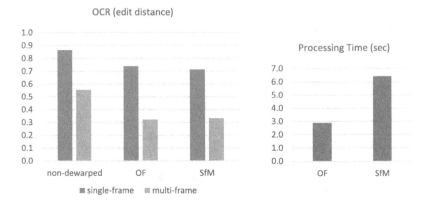

Fig. 9. OCR and processing time results.

For OCR, we use the open-source Tesseract OCR engine [19]. To measure the difference between two text strings, we use edit distance (Levenshtein distance), normalized by dividing by the length of the ground-truth string.

The left and right pages were manually cropped from the images, and each page was processed through the OCR engine. Then the top and bottom boundary text line characters were extracted and the edit distances were computed.

For the single-frame condition, we used the middle frame in the image sequence. For the multi-frame condition, we used the frames at the beginning, middle, and end of the image sequence.

The OCR results show that dewarped was better than non-dewarped, with substantial improvement for multi-frame over single-frame. See Fig. 9. OF and SfM had similar performance for both single-frame and multi-frame. In terms of processing time for computing the 3D structure, OF was much faster than SfM (more than 2x).

6   Conclusion and Future Work

We presented an application to capture page spread images with a mobile phone, and a processing pipeline that uses either OF or SfM to compute the 3D information along with a cylindrical model to perform dewarping. Our preliminary evaluation indicates that OF might be a better choice than SfM since they had similar OCR performance but OF was much faster. This could be important in future systems when the frame images are processed on the mobile phone.

Another aspect that could be improved in the future is to mitigate the motion blur caused by the sweeping motion of the camera when the user takes the sequence of images. This is somewhat noticeable in the images in Fig. 8. One way to address the blur problem is to apply deconvolution algorithms, which is an active area of research (e.g. [21]). Improvements in mobile phone cameras such as faster lens and more reliable autofocus systems will also lessen the blurriness.

Other future work includes automating some of the steps in the pipeline. For example, page frame detection algorithms (e.g. [2]) can be applied to crop the left and right pages from the page spread. Image quality assessment algorithms (e.g. [12]) can be applied to select the frames that are likely to produce the best OCR results.

Acknowledgment. This work was done at FX Palo Alto Laboratory. We thank Michael Cutter and David Lee for helpful discussions.

References

1. Beardsley, P., Zisserman, A., Murray, D.: Sequential updating of projective and affine structure from motion. Intl. J. Comput. Vision **23**(3), 235–259 (1997)
2. Bukhari, S.S., Shafait, F., Breuel, T.M.: Border noise removal of camera-captured document images using page frame detection. In: Iwamura, M., Shafait, F. (eds.) CBDAR 2011. LNCS, vol. 7139, pp. 126–137. Springer, Heidelberg (2012)
3. Bradski, G.: The OpenCV Library. Dr. Dobb's Journal of Software Tools (2000)
4. Brown, M., Seales, W.: Image restoration of arbitrarily warped documents. IEEE TPAMI **26**, 1295–1306 (2004)
5. Brown, M., Tsoi, Y.-C.: Geometric and shading correction for images of printed materials using boundary. IEEE Trans. Image Process. **15**, 1544–1554 (2006)
6. Cao, H., Ding, X., Liu, C.: Rectifying the bound document image captured by the camera: a model based approach. In: Proceedings of ICDAR 2003, pp. 71–75 (2003)
7. Cutter, M., Chiu, P.: Capture and dewarping of page spreads with a handheld compact 3D camera. In: Proceedings of DAS 2012, pp. 205–209 (2012)
8. Fu, B., Wu, M., Li, R., Li, W., Xu, Z., Yang, C.: A model-based book dewarping method using text line detection. In: Proceedings of CBDAR 2007, pp. 63–70 (2007)
9. Liang, J., DeMenthon, D., Doermann, D.: Geometric rectification of camera-captured document images. IEEE TPAMI **30**, 591–605 (2008)
10. Nakajima, N., Iketani, A., Sato, T., Ikeda, S., Kanbara, M., Yokoya, N.: Video mosaicing for document imaging. In: Proceedings of CBDAR 2007, pp. 171–178 (2007)
11. Newman, W., Dance, C., Taylor, A., Taylor, S., Taylor, M., Aldhous, T.: CamWorks: a video-based tool for efficient capture from paper source documents. In: Proceedings of International Conference on Multimedia Computing and Systems, ICMCS 1999, pp 647–653 (1999)
12. Peng, X., Cao, H., Subramanian, K., Prasad, R., Natarajan, P.: Automated image quality assessment for camera-captured OCR. In: Proceedings of ICIP 2011, pp. 2669–2672 (2011)
13. Rother, C., Kolmogorov, V., Blake, A.: Grabcut: interactive foreground extraction using iterated graph cuts. In: Proceedings of Siggraph 2004, pp. 309–314 (2004)
14. Shafait, F., Breuel, T.: Document image dewarping contest, CBDAR 2007
15. Shafait, F., Cutter, M., van Beusekom, J., Bukhari, S., Breuel, T.: Decapod: a flexible, low cost digitization solution for small and medium archives. In: Proceedings of CBDAR 2011, pp. 41–46 (2011)
16. Shi, J., Tomasi, C.: Good features to track. In: Proceedings of CVPR 1994, pp. 593–600 (1994)

17. Szeliski, R.: Computer Vision: Algorithms and Applications. Springer, New York (2010)
18. Taylor, M., Dance, C.: Enhancement of document images from cameras. In: SPIE Conference on Document Recognition V, vol. 3305, 230–241 (1998)
19. TesseractOCR, http://code.google.com/p/tesseract-ocr
20. Triggs, B., McLauchlan, P., Hartley, R., Fitzgibbon, A.: Bundle adjustment a modern synthesis. In: Proceedings of ICCV 1999, pp. 298–372 (1999)
21. Xu, Li, Jia, Jiaya: Two-phase kernel estimation for robust motion deblurring. In: Daniilidis, Kostas, Maragos, Petros, Paragios, Nikos (eds.) ECCV 2010, Part I. LNCS, vol. 6311, pp. 157–170. Springer, Heidelberg (2010)

A Dataset for Quality Assessment of Camera Captured Document Images

Jayant Kumar[✉], Peng Ye, and David Doermann

Institute of Advanced Computer Studies, University of Maryland, College Park, USA
{jayant,pengye,doermann}@umiacs.umd.edu
http://lamp.cfar.umd.edu/media.htm

Abstract. With the proliferation of cameras on mobile devices there is an increased desire to image document pages as an alternative to scanning. However, the quality of captured document images is often lower than its scanned equivalent due to hardware limitations and stability issues. In this context, automatic assessment of the quality of captured images is useful for many applications. Although there has been a lot of work on developing computational methods and creating standard datasets for natural scene image quality assessment, until recently quality estimation of camera captured document images has not been given much attention. One traditional quality indicator for document images is the Optical Character Recognition (OCR) accuracy. In this work, we present a dataset of camera captured document images containing varying levels of focal-blur introduced manually during capture. For each image we obtained the character level OCR accuracy. Our dataset can be used to evaluate methods for predicting OCR quality of captured documents as well as enhancements. In order to make the dataset publicly and freely available, originals from two existing datasets - University of Washington dataset and Tobacco Database were selected. We present a case study with three recent methods for predicting the OCR quality of images on our dataset.

Keywords: Document image quality · Image quality dataset · Sharpness · Optical character recognition

1 Introduction

With the increasing quality of cameras on mobile devices, imaging document pages as an alternative to *scanning* is becoming more feasible ([9,17,18]). However, camera captured document images may suffer from degradations arising from the image acquisition process. One of the most frequently occurring distortions that affects captured image quality is blur. When taking a photo, there are different causes of blur. Figure 1 shows examples of (a) out-of-focus blur, (b) blur due to the motion of camera, and (c) blur due to limited depth of field which occurs when content is at different distances. This is especially apparent in close

M. Iwamura and F. Shafait (Eds.): CBDAR 2013, LNCS 8357, pp. 113–125, 2014.
DOI: 10.1007/978-3-319-05167-3_9, © Springer International Publishing Switzerland 2014

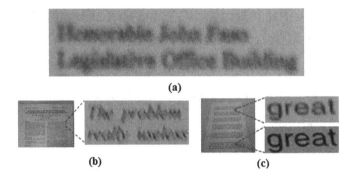

Fig. 1. (a) Out-of-focus blur (b) Motion-blur caused by hand-shake (c) Blur due to limited depth of field when content (characters) are at different distances.

ups and with imaging devices that have a large aperture. Small, high-resolution cameras in smart-phones are more susceptible to these distortions due to their relatively large apertures, and their light-weight and single-hand usage, which make them difficult to hold steady [13].

In the presence of such distortions, the ability to automatically assess the quality of captured images is also becoming increasingly desirable. The required quality of a document image is usually constrained by the applications and usually with respect to human perception or machine readability. An important measure that reflects machine readability is Optical Character Recognition (OCR) accuracy. Predicting OCR accuracy is useful in many different applications. For example, it can be used for selecting the image which will produce the highest OCR accuracy among multiple images of the same document and providing feedback to user in case a re-capture is required. When capturing a document, it is often difficult for a user to determine whether an image is focused on a small mobile screen, so real-time methods for quality estimation can be especially useful [17]. For a large-scale document processing tasks, we can filter out highly degraded document image for which the OCR system would fail. Quality estimation of images has other applications in document analysis tasks including adjusting filters for restoration methods [8], and identifying in-focus and out-of-focus areas of an image [10].

While there has been a lot of work on the creation of standard datasets for scene images ([20, 25, 26]), the quality estimation of camera captured document images has not been given as much attention ([17, 23, 29]). In this work, our goal is to create a dataset of camera captured document images which can be used for the development of quality estimation methods on document images. We have made the dataset publicly and freely available to research community. We selected a set of high-quality document pages from public domain, and used a smart-phone camera for the creation of the dataset. A series of images with varying levels of blur were captured, and OCR accuracies of these images were obtained using three different OCR engines: ABBYY Finereader [4], Tesseract [28] and Omnipage [2]. We evaluated the OCR results of each image against the ground-truth text files.[1] In this work we discuss in detail the creation and characteristics of our dataset. In addition, we present a case-study and discuss results of recent quality estimation methods on our dataset. We hope that our dataset will be useful for researchers working in the area of document image quality assessment.

The remainder of the paper is organized as follows. In Sect. 2 we present the related work on dataset creation for document image quality assessment. We provide details of our dataset in Sect. 3. We then present a case study on OCR quality estimation in Sect. 4 and conclude our paper in Sect. 5.

2   Related Work

In this section, we briefly review existing approaches on the creation of dataset for quality assessment of document images. More specifically we will focus on datasets for estimating OCR quality of document images.

Many datasets for assessing the quality of scanned document images have been discussed ([7,8,27]). One of the early works on predicting OCR accuracy was done by Blando et al. [7]. They used two sets of test data in their experiments. The first set was a subset of ISRI's Sample 2 data base [24] consisting of 460 pages. Each page was digitized at 300 dpi using a Fujitsu M3096M+ scanner. The second set consisted of 200 pages selected from 100 magazines that had the largest paid circulation in the U.S. in 1992. For each magazine, they selected two pages at random and each page was digitized (300 dpi) using a Fujitsu M3096G scanner. The images were binarized using a fixed threshold of 127 out of 255. They used a total of six OCR systems for processing their data sets and collected character accuracy for each image. In their evaluation, each character insertion, deletion, or substitution required to correct the generated OCR text was counted as an error. The character accuracy in their work is defined as:

$$CharacterAccuracy = \frac{n - NumberofErrors}{n} \qquad (1)$$

where n is the total number of characters in the ground-truth text [24].

Cannon et al. [8] focused on the quality of type-written document images and applied it for selecting the optimal restoration approach. They used five quality measures that assess the severity of background speckle, touching characters, and broken characters. They used a dataset of 139 document images with 300 dpi resolution. OmniPage Pro v8.0 was used to perform OCR and the character error rate of the corpus was found to be 20.27 %. They further formed a sub-corpus of 41 documents having OCR character error rates between 20 % and 50 % to perform analysis on highly degraded images. They also created a small corpus of documents spanning a range of gradually decreasing quality by repeatedly photocopying a page from a book (a total of 9 versions). Each successive copy was

---

[1] The OCR results in this paper should in no way be used to compare OCR systems rather only to judge relative performance of each system on the collection.

degraded with background speckle, widened stroke-widths, touching characters and other common attributes of lower quality document images.

Souza *et al.* [27] experimented with a database containing printed documents with a wide variety of font sizes, types and styles. They used a database consisting of 736 documents divided into three sets. Almost all of the images suffer from some type of degradation, such as broken characters, touching characters, salt-and-pepper noise, or the combination of two or more of these problems. Only one printed text line in English was used in all images and none of the images contained any graphics, tables, drawings or underlined text.

Zheng and Kanungo [32] proposed a morphological degradation model based restoration approach for document images. They created a dataset of 100 one-column pages of English Bible that were typeset using LATEX. One additional image was used to estimate pattern distributions. Although the text content of the additional image was different from that of the test images, its font and bigram symbol characteristics were kept similar to the test images. The 100 test images were degraded and categorized into ten groups based on their unique parameter set. They used FineReader 4.0 for OCR and reported reductions in OCR accuracy error rate at the character and word levels ranging from 3.4 % to 41.5 % and from 1.0 % to 20.4 % respectively for different sets of model parameters associated with the degraded images.

Zi [34] presented a document image degradation methodology which incorporated several common types of noise at the page and pixel levels. They developed a system to automatically generate ground truth and degraded images from electronic text. Using their approach, one can produce a complete set of ground truth (text-files and noise free images) which can be used in training or evaluating document analysis systems.

Kumar and Ramakrishnan [16] used a database of 132 annotated multi-script scanned document images comprised of different forms of degradation. They grouped all possible scenarios of the document image degradations to be assigned by a user in the form of a subjective score. Each document image was annotated by 6 users on a scale from 1-5. The dataset is limited to human annotated quality scores and no OCR related analysis was done.

Peng *et al.* [23] proposed an OCR based method which predicts the Normalized Word Error Rate (N-WER) of each document image where a high WER indicates a low image quality. They used a total number of 235 scanned and binarized Arabic text documents from a *Field data set* as original high-quality documents. They captured four degraded images using a digital camera of which two images suffered severe out-of focus blur, and one suffered slight out-of-focus blur. One "clean" image was also captured using an auto-focus feature. The WER for each document was calculated using BBN's OCR engine. The dataset is not available publicly for comparison and analysis of other approaches.

Antonacopoulos *et al.* [6] constructed a dataset consisting of a total of 740 text zone images from a collection of gray scale newspaper images with machine-printed English and Greek text. OCR output from FineReader 9 was used to obtain character level OCR accuracy associated with each image. Ye and

Doermann selected a subset of 521 text zone images which contain more than 30 characters as an experimental set in their work on OCR quality prediction ([29,30]).

Most of the previous work on OCR quality prediction models is limited to scanned document images. The degradations and distortions associated with camera-captured images, however, are very different than scanned images ([23, 30]). The work of Peng *et al.* [23] introduced an approach to create a camera-based document image dataset for OCR quality assessment. But the dataset consisted of only Arabic documents and it is not publicly available. In our work, we have chosen English documents from two publicly available datasets, and made our dataset freely available. Also, unlike [23] which allowed only three levels of degradation based on focus distances, we have captured 6–8 images per document to allow a more continuous OCR quality degradation.

3   Document Image Quality Dataset

In order to make our dataset available, a total of 25 documents from two publicly available data sets - *University of Washington Dataset* [15] and *Tobacco Database* [21] were selected. For each document, 6–8 images were taken from a fixed distance to capture the whole page. The camera was focused at varying distances to generate a series of images with focal blur (as illustrated in Fig. 2). We used a smart-phone[2] with a feature that triggers the camera hardware for focus when the capture button is pressed half-way. Between a fixed minimum and maximum distance to the document, users were instructed to first focus at any distance of their choice. Then a capture was triggered to include the whole document (including borders). The focus distance was decided by user and we did not calibrate distances across different captures. Other conditions including lighting and place of capture were kept the same for all documents. One of the shots taken was sharp, i.e., focus and capture is done at the same (fixed) distance. A total of 25 such sets, each consisting of 6–8 high-resolution images (dimension: 3264×1840) were created using an Android phone with an 8 megapixel camera. The dataset has a total of 175 images. Figure 3 shows three sample images from our dataset, and the corresponding OCR accuracy.

We used three popular OCR engines to process the images: ABBYY FineReader 10 [4], Tesseract [28] and Omnipage [2]. We used the batch-mode default settings and saved the generated text-files in the plain text format. We obtained character level accuracy for each captured image in our dataset using the ISRI-OCR evaluation tool [3]. The tool's program **accuracy** generates a character accuracy report when a *correct* and *OCRed* file is given. We used the program in the case-insensitive mode. A character accuracy report consists of six sections. The first section specifies the number of characters in the ground truth, the number of errors made by the OCR engine, and the character accuracy (as percentage).

---

[2] Motorola DroidX with Android.

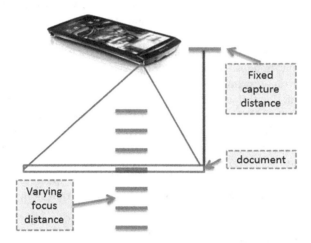

Fig. 2. Creation of our *Document Image Quality Assessment* dataset. Images were captured from a fixed distance to include the whole page (including borders). The camera was focused at varying distances to generate a series of images with focal blur.

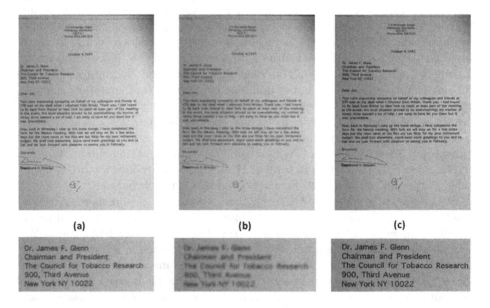

Fig. 3. Sample images from our dataset showing three levels of focus-blur. The sharpness of each image is different due to variation in focal-distance. The OCR accuracy for three images are (a) 96 % (b) 13 % (c) 99 %. The second row shows a portion of text content from each image.

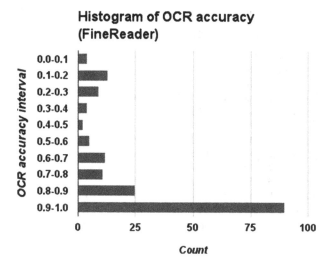

Fig. 4. Histogram of OCR accuracies in our dataset using Finereader. The vertical axis shows the interval of OCR accuracy and horizontal axis shows the number of documents.

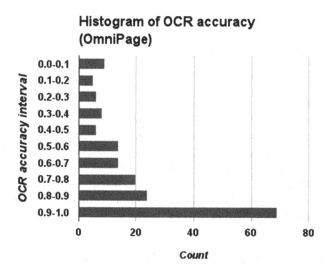

Fig. 5. Histogram of OCR accuracies in our dataset using OmniPage. The vertical axis shows the interval of OCR accuracy and horizontal axis shows the number of documents.

In our first release we provide the following for download: (1) 25 sets of camera captured images each containing 6–8 images of a particular document, (2) three OCR text files corresponding to three OCR engines used for each captured image and ground-truth text file, and (3) OCR accuracies associated with each

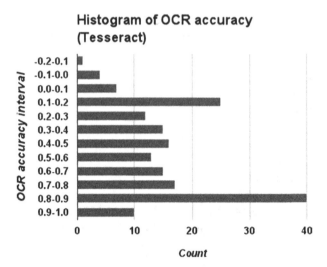

Fig. 6. Histogram of OCR accuracies in our dataset using Tesseract. The vertical axis shows the interval of OCR accuracy and horizontal axis shows the number of documents. Negative OCR accuracy may occur in highly degraded document image when OCR engine treats some non-text regions (e.g. figures) as text regions and generates text for these regions.

captured image. Figure 4, 5 and 6 shows the histograms of OCR accuracies in our dataset using FineReader, OmniPage and Tesseract respectively. As observed in Fig. 6, some documents have negative OCR accuracy when the number of errors is more than the number of characters (Eq. 1). The number of characters is computed using the groundtruth file and errors are defined as the actual edit operations (character insertions, substitutions, and deletions). Negative OCR accuracy may occur in highly degraded document image when OCR engine treats some non-text regions (e.g. logo, figures) as text regions and generates text for these regions. In that case, extra deletion operations are required and the number of errors may be larger than the original number of characters in the groundtruth. Our dataset is publicly available for download at [19].

4   Case Study: OCR Quality Prediction

In this section we discuss results of three recent methods on quality assessment on our dataset. We used the OCR text output from ABBYY's *FineReader* in this case study. FineReader is a widely used OCR software [5] and provides the best performance among the three OCR softwares we have tested on our dataset. In this case study, we limit the evaluation to OCR text obtained using FineReader.

4.1 Methods and Evaluation

We evaluated two unsupervised sharpness estimation methods for OCR quality prediction on our dataset. These methods were developed for estimating human-perceived sharpness of images. We are interested in evaluating whether the *sharpness scores* computed by these methods are good indicator of OCR quality. Additionally, we tested a supervised approach of Ye and Doermann [29] based on feature learning which showed promising results on a set of scanned gray-scale document images. The three methods selected are as follows:

1. **Q:** Zhu *et al.* [33] proposed a no-reference sharpness metric (Q) based on singular value decomposition (SVD) of the local image gradient matrix. The method was shown to perform well on the parametrization of an image restoration algorithm.

2. **△DOM:** Kumar *et al.* [17] presented a fast sharpness estimation approach (△DOM) for smart-phone based document images where degradation is common due to defocus or camera-motion. Their experiments with a corpus of document images that they collected and labeled using workers from Amazon's *Mechanical Turk* show that the performance of their method is better than state-of-the-art perceptually-based models ([14, 22]).

3. **CORNIA:** Ye and Doermann [29] proposed an unsupervised feature learning framework to learn effective features directly from the training data for predicting OCR accuracy of gray-scale document images. The first step in their approach involves extracting raw-image-patches from a set of unlabeled images to learn a dictionary using a clustering method. For the OCR quality prediction on a given image, a set of raw-image patches are extracted and encoded using the learned dictionary based on soft-assignment encoding with max pooling. In the last step, Support Vector Regressor (SVR) [11] is used to learn a mapping from the image features to an image quality score. By learning a compact set of filters CORNIA was shown to perform real-time quality estimation ([30, 31]).

We used the MATLAB implementation provided by each of the authors for evaluation. First two methods do not require an explicit training phase, and the parameters were tuned based on cross-validation on different sets. For CORNIA, we used a 25-fold cross-validation scheme in which images from 24 sets were used for training and the remaining set was used for testing. This procedure was repeated for all 25 sets.

We used two metrics for evaluating the performance of different systems. The first was the Spearman Rank Order Correlation Coefficient (SROCC) to measure how well the rank assigned by each method correlate with the ranked OCR accuracies. The second was the Pearson (or Linear) Correlation Coefficient (LCC) to measure the linear dependence between scores and OCR accuracy. While SROCC is a monotonicity measure of a prediction model the second metric LCC measures the strength and the direction of a linear relationship([1, 12]).

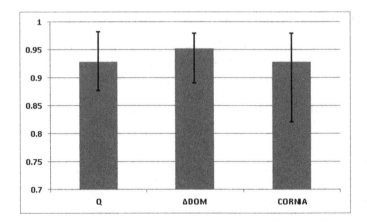

Fig. 7. Median Spearman rank correlation for 25 sets in our dataset. The upper and lower end of line segments represent the 75th and 25th percentile respectively.

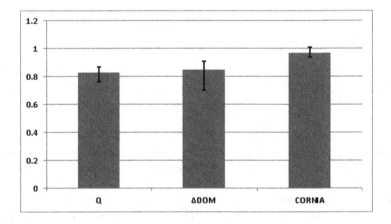

Fig. 8. Median Pearson linear correlation for 25 sets in our dataset. The upper and lower end of line segments represent the 75th and 25th percentile respectively

4.2 Results and Discussion

Figure 7 summarizes the results of Spearman rank correlation (SROCC) values for 25 sets in our data. We computed the SROCC for each set using the scores computed by each method against the OCR accuracy. The top of the bars in Fig. 7 indicate observation *median* and the line segments represent the 75th and 25th percentile. Table 1 provides exact correlation scores for comparisons.

Of the three methods, $\triangle DoM$ performed consistently well on all the sets, while Q and CORNIA showed relatively higher variation in results on different sets. A higher SROCC value indicates the method's ability to rank images for a particular document, and can be used to select the image with best OCR accuracy.

Table 1. Spearman rank correlation and Pearson linear correlation between OCR accuracies and quality scores of three tested methods.

Spearman Rank Correlation			
	Q	CORNIA	$\triangle$DoM
Median	0.9370	0.9286	**0.9370**
25th Percentile	0.8850	0.8214	0.8850
75th Percentile	0.9910	0.9799	0.9910
Pearson Linear Correlation			
Median	0.8271	**0.9747**	0.8488
25th Percentile	0.7631	0.9447	0.7001
75th Percentile	0.8681	0.9822	0.9061

Figure 8 shows the box-plot for Pearson correlation scores for 25 sets. Similar to previous plot, the bar shows the 75th and 25th percentile of scores. A good correlation score is needed for applications such as determining whether a captured image is good enough to keep or should be retaken. CORNIA performed better than other two approaches on modeling the linear relationship between two variables. When the goal of quality estimation is to predict the true quality score of images with different underlying content, CORNIA (or other supervised methods) usually outperforms unsupervised approaches.

5 Conclusion

We have created a dataset for evaluating document image quality assessment approaches. Our dataset and related data is publicly and freely available for download. To the best of our knowledge, this is the first publicly available dataset for camera captured document image quality assessment. This dataset will be useful to researchers working on the purposive evaluation of quality estimation methods for predicting the OCR quality of document images. The dataset has a total of 525 (175 $\times$ 3) OCR-text files from three popular OCR engines. Furthermore, we also obtained character level accuracy for each OCR-text file. In future versions of this dataset, we would like to obtain human-perceived quality of each image. We also plan to add images representing other distortions such as low-light and motion-blur to our dataset.

We also presented results of three recent methods on estimating the OCR quality of images based on output obtained from FineReader. Using two different evaluation measures we compared and discussed the advantages of three quality estimation approaches. Our case study showed that $\triangle$DOM is effective for ranking images based on OCR quality and CORNIA is effective for obtaining the true quality scores of document images.

Acknowledgments. We would like to thank Steven Dang for running Tesseract on our images. We would also like to thank Francine Chen and anonymous reviewers for their comments on improving the quality of this work. The partial support of this research by DARPA through BBN/DARPA Award HR0011-08-C-0004 under subcontract 9500009235, and the US Government through NSF Award IIS-0812111 is gratefully acknowledged.

References

1. Spearman's rank correlation coefficient. http://en.wikipedia.org/wiki/Spearman's_rank_correlation_coefficient
2. Omnipage professional version 18.0. http://www.nuance.com/for-business/by-product/omnipage/index.htm (2011)
3. ISRI-OCR evaluation tool: Code and data to evaluate OCR accuracy, originally from UNLV/ISRI. http://code.google.com/p/isri-ocr-evaluation-tools/ January 2010
4. ABBYY Finereader 10 Professional Edition, build 10.0.102.74 (2009)
5. ABBYY finereader 8.0 professional edition, September 2005
6. Antonacopoulos, A., Clausner, C., Papadopoulos, C., Pletschacher, S.: Historical document layout analysis competition. In: International Conference on Document Analysis and Recognition (ICDAR), pp. 1516–1520 (2011)
7. Blando, L., Kanai, J., Nartker, T.: Prediction of OCR accuracy using simple image features. In: International Conference on Document Analysis and Recognition, vol. 1, pp. 319–322 (1995)
8. Cannon, M., Hochberg, J., Kelly, P.: Quality assessment and restoration of typewritten document images. Int. J. Doc. Anal. Recogn. **2**(2–3), 80–89 (1999)
9. Chen, F., Carter, S., Denoue, L., Kumar, J.: SmartDCap: semi-automatic capture of higher quality document images from a smartphone. In: International Conference on Intelligent User Interfaces (IUI), pp. 287–296 (2013)
10. Chung, Y.C., Wang, J.M., Bailey, R., Chen, S.W., Chang, S.L.: A non-parametric blur measure based on edge analysis for image processing applications. In: IEEE Conference on Cybernetics and Intelligent Systems, vol. 1, pp. 356–360 (2004)
11. Drucker, H., Burges, C.J.C., Kaufman, L., Smola, A., Vapnik, V.: Support vector regression machines. In: Mozer, M., Jordan, M., Petsche, T. (eds.) Advances in Neural Information Processing Systems, vol. 9, pp. 155–161. MIT Press, Cambridge (1997)
12. Edwards, A.L.: The correlation coefficient: An Introduction to Linear Regression and Correlation, pp. 33–46. W. H. Freeman, San Francisco (1976)
13. Fergus, R., Singh, B., Hertzmann, A., Roweis, S.T., Freeman, W.T.: Removing camera shake from a single photograph. ACM Trans. Graph. **25**, 787–794 (2006)
14. Ferzli, R., Karam, L.: A no-reference objective image sharpness metric based on the notion of just noticeable blur (JNB). IEEE Trans. Image Process. **18**, 717–728 (2009)
15. Guyon, I., Haralick, R.M., Hull, J.J., Phillips, I.T.: Data sets for OCR and document image understanding research. In: Proceedings of the SPIE - Document Recognition IV, pp. 779–799. World Scientific (1997)
16. Kumar, D., Ramakrishnan, A.: Quad: quality assessment of documents. In: International Workshop on Camera based Document Analysis and Recognition, pp. 79–84 (2011)
17. Kumar, J., Chen, F., Doermann, D.: Sharpness estimation of document and scene images. In: International Conference on Pattern Recognition (ICPR), pp. 3292–3295 (2012)

18. Kumar, J., Bala, R., Ding, H., Emmett, P.: Mobile video capture of multi-page documents. In: IEEE International Workshop on Mobile Vision (IWMV), pp. 35–40 (2013)
19. Kumar, J., Ye, P., Doermann, D.: DIQA: document image quality assessment datasets. In: Language and Media Processing Laboratory (2013). http://lampsrv02.umiacs.umd.edu/projdb/project.php?id=73
20. Larson, E.C., Chandler, D.M.: Most apparent distortion: full-reference image quality assessment and the role of strategy. J. Electron. Imaging 19(1), 1–21 (2010)
21. Lewis, D., Agam, G., Argamon, S., Frieder, O., Grossman, D., Heard, J.: Building a test collection for complex document information processing. In: International ACM SIGIR Conference on Research and Development in Information Retrieval, pp. 665–666. ACM (2006)
22. Narvekar, N., Karam, L.: A no-reference image blur metric based on the cumulative probability of blur detection (CPBD). IEEE Trans. Image Process. 20(9), 2678–2683 (2011)
23. Peng, X., Cao, H., Subramanian, K., Prasad, R., Natarajan, P.: Automated image quality assessment for camera-captured OCR. In: IEEE International Conference on Image Processing (ICIP), pp. 2621–2624 (2011)
24. Rice, S.V., Kanai, J., Nartker, T.A.: The third annual test of OCR accuracy. TR 94–03 ISRI. University of Nevada, Las Vegas (1994)
25. Sheikh, H.R., Wang, Z., Cormack, L., Bovik, A.C.: Live image quality assessment database release 2 (2006). http://live.ece.utexas.edu/research/quality
26. Sheikh, H., Sabir, M., Bovik, A.: A statistical evaluation of recent full reference image quality assessment algorithms. IEEE Trans. Image Process. 15(11), 3440–3451 (2006)
27. Souza, A., Cheriet, M., Naoi, S., Suen, C.: Automatic filter selection using image quality assessment. In: International Conference on Document Analysis and Recognition, pp. 508–512 (2003)
28. Tesseract-OCR: An OCR engine that was developed at HP Labs between 1985 and 1995 and now at Google. https://code.google.com/p/tesseract-ocr/ (2012)
29. Ye, P., Doermann, D.: Learning features for predicting OCR accuracy. In: International Conference on Pattern Recognition (ICPR), pp. 3204–3207 (2012)
30. Garibotto, G., et al.: White paper on industrial applications of computer vision and pattern recognition. In: Petrosino, Al (ed.) ICIAP 2013, Part II. LNCS, vol. 8157, pp. 721–730. Springer, Heidelberg (2013)
31. Ye, P., Kumar, J., Kang, L., Doermann, D.: Unsupervised feature learning framework for no-reference image quality assessment. In: International Conference on Computer Vision and Pattern Recognition (CVPR 2012), pp. 1098–1105 (2012)
32. Zheng, Q., Kanungo, T.: Morphological degradation models and their use in document image restoration. Technical Report LAMP-TR-065, CS-TR-4218, CAR-TR-962, University of Maryland, College Park, February 2001
33. Zhu, X., Milanfar, P.: Automatic parameter selection for denoising algorithms using a no-reference measure of image content. IEEE Trans. Image Process. 19(12), 3116–3132 (2010)
34. Zi, G.: GroundTruth generation and document image degradation. Technical Report LAMP-TR-121, CAR-TR-1008, CS-TR-4699, UMIACS-TR-2005-08, University of Maryland, College Park, May 2005

A Morphology-Based Border Noise Removal Method for Camera-Captured Label Images

Mengyang Liu[✉], Chongshou Li, Wenbin Zhu, and Andrew Lim

Department of Management Sciences, College of Business,
City University of Hong Kong, Kowloon Tong, Hong Kong SAR
mengyliu2-c@my.cityu.edu.hk

Abstract. Printed labels are widely used in our life to track items, especially in logistics management. If item information on a label could be recognized automatically, the efficiency of the logistics would be greatly improved. However, some particular properties of label images make them difficult for off-the-shelf optical character recognition (OCR) system to recognize directly. To prepare the label images for OCR, border noise removal is an important step. With text region only, the resulting image would be easier for OCR to read. In this paper, we propose a simple and effective approach to remove border noise in textile label images. Border noise in those label images is more complex than that in conventional document images. Our solution consists of four parts: label boundary detection, label blank region extraction, holes filling and border noise deletion. The experiment shows that the proposed method yields satisfactory performance.

1 Introduction

In our daily life, printed labels are around us almost everywhere. We read it, and get information about the item from it. In logistics management, printed labels also play important roles. They carry and share information among various companies along a supply chain. The information on the labels is extracted in process such as stock taking. The productivity will be improved significantly, if the information extraction can be carried out automatically. In practice, automatic label recognition system is often utilized, for example barcode. However, barcode has its limitations. It is not human readable. The use of barcode requires coordination of various parties involved: at least their IT system must be able to communicate with each other to share barcode data. Many small and medium business units today still lack the necessary IT systems to utilize barcode. Traditional printed labels are still in wide use and will continue to be widely used. Figure 1(b) illustrates a typical printed label that is currently used in a textile warehouse in Hong Kong (Fig. 1(a)).

Smart phones nowadays coming with a camera as standard component are widely available. The processing power of such devices are comparable to old desktops. With the maturity of Optical Character Recognition (OCR) technology in recent years, it is possible to develop an OCR based automatic label

M. Iwamura and F. Shafait (Eds.): CBDAR 2013, LNCS 8357, pp. 126–138, 2014.
DOI: 10.1007/978-3-319-05167-3_10, © Springer International Publishing Switzerland 2014

(a)   (b)

Fig. 1. (a) A textile warehouse in Hong Kong. (b) A typical printed label without barcode.

recognition software solution and use the camera on the smartphones to replace the barcode scanners.

However, off-the-shelf OCR softwares are often optimized for scanned document images which usually contain little noise and distortion. Whereas, images taken by a phone from a typical logistics operation are usually highly distorted and contaminated with noise. Distortions are introduced by the perspective projections and nonplanar surfaces onto which the labels are pasted. Noises are usually introduced by dusts, uneven lighting, complex patterns of the materials onto which the labels are pasted. Figure 2 illustrates a few typical label images from production environments. It is crucial to remove as much noise as possible before passing the image to OCR module to achieve sufficient accuracy for production use. In this study, we focus on removing the border noise, which refers to the noise in the margin of an image. More specifically we concentrate on labels that are pasted on rolls of textiles. The border noise in this case is mainly introduced by textile surrounding the label.

Border noise removal is well-known and has been studied in the field of document image analysis. Reference [1] has carefully analyzed the effect of border noise on page segmentation. Briefly speaking, this research can be divided into

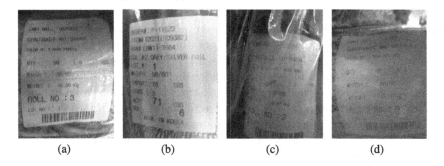

(a)   (b)   (c)   (d)

Fig. 2. Examples of label image.

two classes: border noise removal for scanned document images and border noise removal for camera-captured document images. For scanned document images, border noise removal has received enough attention. Essentially, two established main approaches blanket the literature ([2]). One is detecting and removing noise region while the other is directly identifying actual content region. However, limited work has been reported for removing border noise in images captured with a camera.

Studies concentrating on detection and removal of noisy region include [3–7]. Reference [3] proposes several heuristic procedures with empirical thresholds to detect borders of text regions. This work relies on the assumption that borders are very close to image edges and there is a large white areas between text borders and image edges. Obviously, this assumption can not hold in our application (see Fig. 2). References [4,6] focus only on removing non-textual border noise. Removing both textual and non-textual noise is addressed by [5,7]. These methods can be effective due to several underlying assumptions for scanned document images. For example, the shape of border noises are regular and parallel to image edges; there exists fat black non-textual noise region. In the case of label images, the border noise are more complex and versatile due to the various textiles rolls which the labels are pasted on.

There also exist researches directly identifying content region in scanned document images. Reference [8] proposes to find the optimal page frame of structured documents (journal articles, books, magazines) by a geometric matching algorithm. The method reported by [9] is to detect the optimal page frames of double-page scanned document images and divide into two pages without border noise.

A few studies of border noise removal for camera-captured images have been found in the existing literature ([2,8,10]). Reference [10] proposes a method based on projection profile. The assumption behind it is that there exists blank region between texts region and border noise. Reference [8] applied their page frame detection algorithm to camera-captured images. This method relies on the observation that distortion only happens on the top and bottom of the page frame. The proposed page frame model can still be used to estimate left and right borders and remove noise in the corresponding sides of images. In our application, distortion occurs not only on the top and bottom of label region but also on the left and right sides of label region (see Fig. 11(e)). Recently, [2] proposes to use information of text and non-text region to identify page frame and remove border noise. An important step in this method is segmentation of text and non-text region. In our label images, the size of non-text element is not significantly larger than that of text element. The multiresolution morphology based segmentation method with threshold reduction used by [2] might not be suitable to effectively separate text and non-text region in our label images.

In this paper, we propose a border noise removal approach for label images. Labels and textiles are quite different materials. We can expect a noticeable difference across the boundary of the label. Our idea is that we first try to detect the border of a label through this difference in Sect. 2. We then identify the connected component inside the label boundary, which is most likely the blank region of the label in Sect. 3. After that, we fill the holes inside the blank region, which are most likely caused by texts on the label in Sect. 4. At last, we delete the border noise in Sect. 5. Through these four steps we can properly remove most of the border noise. We report some experimental results in Sect. 6 and discuss this study in Sect. 7.

2   Detect Label Boundary

When detecting label boundary, all the pixels on the boundary should be highlighted and the label should not be separated.

To detect the boundary of the label, one method is to extract the label region, which is exactly the objective of border noise removal, based on the difference between label and the background. Commonly used feature is gray value of pixels. However, binary method such as Otsu's global thresholding method [11] is incapable. Figure 3(a) shows an example. The white pixels in the resulting binary image represents only part of the label.

Another choice is to detect the boundary of the label directly. Edge detection method seems to be a good choice. However, methods such as Canny's edge detector ([12]) would introduce too much noise (Fig. 3(b)). As a result, the label region would be divided into different parts and we can't get a complete image of the label if using the same idea as our proposed method. In Fig. 3(c), the label is divided into several parts, and the parts that connect with the border noise are showed as background. Therefore, we need a method that could extract the whole label boundary without separating the labels to identify the entire label region in the next few steps.

Observe that labels and textiles are very different materials. Within a small region in the same material, we can expect certain form of regularity, such as color, texture, and reflection. Therefore we can expect a form of local similarity, that is, image pixels in a small region of the same material look similar. Such similarity does not hold at global level, that is, two regions far away from each

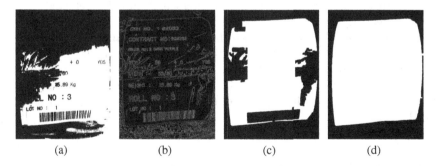

(a)   (b)   (c)   (d)

Fig. 3. Comparison of different methods. (a) Applying Otsu' method in Fig. 2(a). (b) Edge image of Fig. 2(a) constructed by using Canny's detector. (c) Consequence of using edge image instead of contrast image in the proposed method. (d) Final label region of Fig. 2(a) extracted by the proposed method.

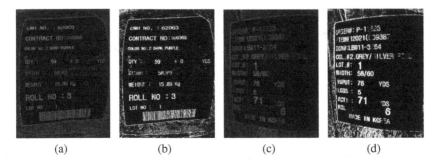

(a)   (b)   (c)   (d)

Fig. 4. (a) Contrast image of Fig. 2(a). (b) Binary contrast image. (c) Contrast image of Fig. 2(b). (d) Binary contrast image.

other will be quite different even if they are from the same material. However, local similarity is sufficient for us to detect the border of label. Consider a small section of a border and two small regions across the border. One region contains pixels representing the textile and the other region contains pixels representing the label. We can expect pixels within a region are quite similar and pixels across the regions are quite different due to difference in material. Therefore, if we construct a contrast image, we can expect the contrast in a region is low (since the pixels are similar) and the contrast of pixels on the border is high. Similar principles have been employed by [13].

In addition, labels usually contain a substantial area of blank region. Pixels in the blank region are very similar and will have lower contrast even with shadow on it. When we transform the contrast image into a binary image, we can expect the blank region of the label to be a continues region inside the border.

We construct the contrast image using the same formula as Formula 2 in [14, Formula 2]. The resulting contrast image is showed in Fig. 4(a).

We transform the resulting contrast image into a binary image using Otsu's global thresholding method ([11]). Most of the pixels on the boundary of the label will be detected as high contrast pixels. However, some pixels on the boundary of a pattern on the textiles or text characters may also be detected as high contrast pixels. Figure 4(b) illustrates an example, where high contrast pixels are displayed in white. We can clearly see the border of the label. The white pixels inside the borders are mostly from the text characters on the label. The white pixels outside the borders may introduce border noises that we must remove. At this stage, we still cannot isolate pixels that represent borders from noises. We will show in next Section that it is easier to identify the blank region in the label instead. With some care, the boundary of the blank region in the label will help us identify the border of the label.

3   Extract Blank Region of a Label

The blank region in a label forms a connected component in the binary contrast image constructed in previous step. Although connected component can be

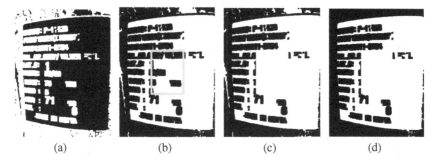

(a)   (b)   (c)   (d)

Fig. 5. Label blank region extraction process applied to Fig. 4(d). (a) after dilating, (b) reversing the image, (c) set all pixels in the yellow rectangle to white, (d) connected component containing the yellow window (Color figure online)

easily detected, care must be taken to ensure that a connect component dose not include extra regions. When there are gaps on the border, the connected component would expand beyond the label region, as illustrated in Fig. 12. A simple fix is to apply morphological operation to remove as many gaps on the border as possible.

The remaining challenge is that there are usually more than one connected components in the binary contrast image. It is tempting to assert that the largest connected component is the blank region. This assertion is true if the label occupies a substantial portion of the entire image. In cases, where images are taken from a long distance, a label is only a very small portion in the image. In such cases, we need some additional information to help identify the blank region. Fortunately in our application, end users can easily provide such information. We can provide a visual cue in the user interface and ask user to make sure the yellow window in the center of an image is fully inside the border of a label, see Fig. 5(b).

We extract blank region in four steps. We use the binary contrast image in Fig. 4(d) to illustrate our process. The result of each step is given in Fig. 5.

1. Dilate the binary contrast image. This will connect nearby edges and therefore fill up small gaps on the border of the label, see Fig. 5(a). Using larger window size will allow us to fill larger gaps on the border. However when the window size is too large, we may incidentally merge text characters that are close to borders into the border and therefore lose such characters. Experiment shows that flat structuring element of dimensions 10×10 is proper.
2. Reverse the binary image so that the white pixels represent the low contrast pixels, which include the blank part of the label and low contrast pixels in the margin (Fig. 5(b)).
3. Set all pixels in the yellow window in Fig. 5(b) to white color. The result is shown in Fig. 5(c).
4. Start from a pixel in the yellow window, find all white color pixels connected with it using the "bwlabel" function described in [15, pp. 360–362]. The white pixels in Fig. 5(d) shows the extracted component. Note that the noise outside the border are not included in the extracted component.

Observe that the extracted label blank region contains many holes, see Fig. 5(d). Those holes are most likely caused by texts on the label. Therefore, in the next step, we would like to fill the holes, which will allow us to extract the entire label from the image.

4   Patch Holes Inside Label Blank Region

There are two types of holes: (1) holes in the middle of the blank region that are completely surrounded by the blank region; (2) holes at the border of the label. The first type of holes can be easily filled up using hole filling operations described in [15, pp. 365–366]. Figure 6(a) shows the result.

To fill the holes that are at the border of a label, we apply the closing operation described in [15, pp. 347–350]. The structuring element of the operation must be able to eliminate the holes but not bring too much new noise. To reach that goal, we choose the flat structuring element that is slightly larger than the average size of a text character on the label. Note that components inside the type 1 holes are most likely caused by text characters on the label. They are either a complete character or a part of a character. Therefore we can use their average size to estimate the size of a text character.

Let h_i be the height of the ith component in type 1 holes. Let H be the height of entire image. We propose four different formulas to estimate the average height of a character:

$$
\begin{aligned}
c_1 &= \text{mode of } \{h_i\} \\
c_2 &= \text{mode of } \{h_i \mid h_i \neq c_1\} \\
c_3 &= \text{mode of } \{h_i \mid h_i \neq c_2 \text{ and } h_i \neq c_1\} \\
c_4 &= \text{mean of } \{h_i \mid 10 < h_i < H/10\}
\end{aligned}
\tag{1}
$$

Note that when computing c_4, we filtered out components that are unlikely to be a character. Components that are too small, say smaller than 10 pixels are

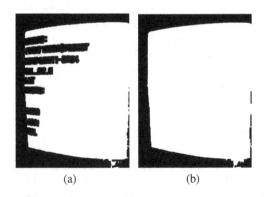

(a)   (b)

Fig. 6. Filling holes in the blank region. (a) after fill the holes inside; (b) after closing

usually caused by noises such as dirts. Components that are too large, say larger than one tenth of the height of entire image are unlikely to be a character either.

The unexpected broken of the characters, or the existence of some small components such as dots often leads to underestimate of the true size of a character. To compensate for such underestimation, we take the largest among $c_1, \cdots, c_4$ as our final estimate of the average height of a character and denote it by h. The window size for the closing operation is set to $1.5h \times 3$. The result of closing operation is shown in Fig. 6(b). As we can see, the white region in the resulting binary image is very similar to the actual label. In the next step, we would use this binary image as mask to remove border noise.

5   Delete Border Noise

Let $B(x, y)$ denote the value of pixel (x, y) in the binary image we get in the previous step. Let $f(x, y)$ denote the value of pixel (x, y) in the new image after removing border noise. Let $f_0(x, y)$ denote the value of pixel (x, y) in the binary image of original label image. The binary images are constructed using method in [13]. The corresponding contrast image is constructed using formula in [14, Formula 2]. Then the process to remove the border noise is as follows.

$$f(x, y) = \begin{cases} 0 & \text{if } B(x, y) = 0 \\ f_0(x, y) & \text{if } B(x, y) = 1 \end{cases} \quad (2)$$

To better illustrate the result, we implement the same process in the original gray-scale image. The result for Fig. 2(b) is shown in Fig. 7.

In some cases, the extracted label region may contain more than the actual label, and it seems that the border noise is not completely removed, as showed in Fig. 8(b). The reason of the extra region is that, although in Sect. 3, we have dilated the image to connected unexpected broken boundary, there may still exist large gaps that are not able to be eliminated by the dilating operation. However,

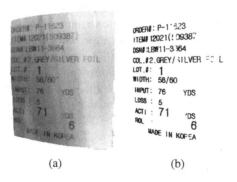

(a)   (b)

Fig. 7. Border noise removing result for Fig. 2(b). (a) gray-scale image after removing border noise. (b) binary image after removing border noise.

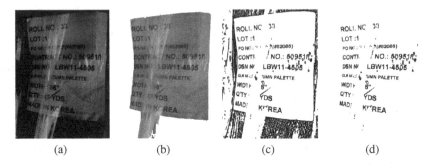

| (a) | (b) | (c) | (d) |

Fig. 8. (a) original image. (b) gray-scale image after removing border noise. (c) binary image for the original image. (d) binary image after removing border noise.

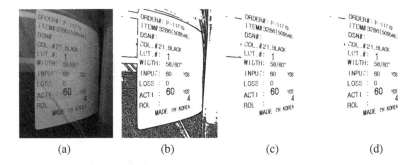

| (a) | (b) | (c) | (d) |

Fig. 9. (a) original image, (b) binary image, (c) binary image after removing border noise, (d) binary image after removing components with abnormal size (noise)

since those extra region is caused by low contrast pixels touching the gaps, most of them would be transformed to 0 in the final binary image. Therefore, the binary image (Fig. 8(d)) would not be affected too much by those extra region. In most cases, the image fed to OCR must be in binary form, which means the final recognition result would not be affected by the extra region.

But there still exists cases where border noise are not able to be removed in the binary image. When the extra region contains noise that does not touch the boundary of the image, the isolated noise would be kept after filling holes in Sect. 4, as showed in Fig. 9(c). Generally speaking, most of those noises are of abnormal size compared with the characters. Therefore, we apply a filter to remove them by size. The average height of text characters (h) has been estimated in Sect. 4. We estimate the average width of text characters, which we denote as w, in a similar fashion. Then, we delete the components with a height more than $3 * h$ or a width more than $5 * w$. The result is shown in Fig. 9(d).

6   Experimental Results

To analyze the accuracy of the proposed approach, we conducted experiment on our date set, which consists of images of different labels pasted on various textile products. The resolutions of all the images are either 612 × 816 or 750 × 1000. We get these images by taking pictures of labels in textile warehouse with an iphone 4S and Samsung smart phone with Andriod system.

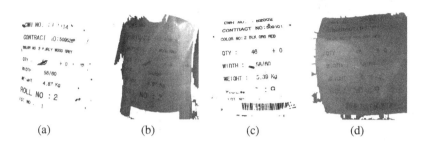

(a)   (b)   (c)   (d)

Fig. 10. Results for Fig. 2(c) and (d). (a), (b) final binary image and gray image for Fig. 2(a); (c), (d) final binary image and gray image for Fig. 2(d).

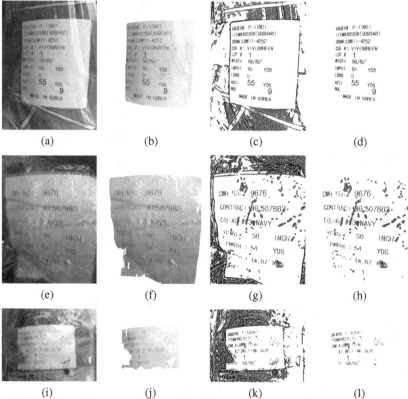

(a)   (b)   (c)   (d)

(e)   (f)   (g)   (h)

(i)   (j)   (k)   (l)

Fig. 11. The first column contains original images. The second column contains gray-scale images after removing border noise for corresponding images in the first column. The third column contains binary images for the original images. The last column shows the binary images after removing border noise.

Fig. 12. The first column contains original images. The second column contains gray-scale images after removing border noise for corresponding images in the first column. The third column contains binary images for the original images. The last column shows the binary images after removing border noise.

Generally speaking, the proposed approach showed to be effective and efficient. Figure 10 shows result for Fig. 2(c) and (d). Because of the plastic film, the whole images are gray and not clear. The label region are almost successfully detected. Figure 10(b) and (d) show the result of binary images.

Figure 11 shows another 3 examples. Border noise in these images have different patterns, colors and texture. The border noise of label in Fig. 11(a) is removed perfectly as showed in Fig. 11(b) and (d). The labels in Fig. 11(e) and (i) are of worse condition. They are skewed, spotted or with obvious plastic film. But the detected label regions of these two cases are almost the same with the true label region. Although Fig. 11(h) still has noise, the border noise has been removed successfully.

Another 3 examples are displayed in Fig. 12. In those cases, the identified label regions have extra area compared with the true region. Those are caused by the low contrast pixels that locates on the label boundary (pixels on the upper border in Fig. 12(c), lower border in Fig. 12(g) and lower border in Fig. 12(k)). But the resulting binary images do not contain too much border noise, see Fig. 12(d), (h) and (l).

7 Discussion

In this paper, a novel border noise removal approach is proposed. It contains four main procedures. They are label boundary detection, blank region extraction, holes filling and border noise deletion. Our experiment shows that it works well in practice.

Although the proposed approach is designed for textile label recognition system, it would also works well with label and document images in different cases.

When applied in different cases, instead of image contrast, other techniques could be applied to detect the boundary of the text region, i.e., the physical limit of labels or pages. The condition of the scanning document image is always better than the label images. Therefore, binarization method could be adapted. If the edge is not distinct, more sensitive methods such as Canny Edge Detector could be used. Once the edge of text region is detected, the procedure in Sects. 3 and 4 could be applied.

One constraint for our proposed method is that the central region of the label should be part of the text region (label). However, this constraint could be relaxed. In more general cases, if the time complexity of the method is not a major issue, we could apply some text detection method to locate the text, and thus locating a start region.

References

1. Shafait, F., Breuel, T.M.: The effect of border noise on the performance of projection-based page segmentation methods. IEEE Trans. Pattern Anal. Mach. Intell. **33**(4), 846–851 (2011)
2. Bukhari, S.S., Shafait, F., Breuel, T.M.: Border noise removal of camera-captured document images using page frame detection. In: Iwamura, M., Shafait, F. (eds.) CBDAR 2011. LNCS, vol. 7139, pp. 126–137. Springer, Heidelberg (2012)
3. Le, D.X., Thoma, G.R., Wechsler, H.: Automated borders detection and adaptive segmentation for binary document images. In: Proceedings of the 13th International Conference on Pattern Recognition, 1996, vol. 3, pp. 737–741. IEEE (1996)
4. Fan, K.C., Wang, Y.K., Lay, T.R.: Marginal noise removal of document images. Pattern Recogn. **35**(11), 2593–2611 (2002)
5. Cinque, L., Levialdi, S., Lombardi, L., Tanimoto, S.: Segmentation of page images having artifacts of photocopying and scanning. Pattern Recogn. **35**(5), 1167–1177 (2002)

6. Ávila, B.T., Lins, R.D.: Efficient removal of noisy borders from monochromatic documents. In: Campilho, A.C., Kamel, M.S. (eds.) ICIAR 2004. LNCS, vol. 3212, pp. 249–256. Springer, Heidelberg (2004)

7. Shafait, F., Breuel, T.M.: A simple and effective approach for border noise removal from document images. In: IEEE 13th International Multitopic Conference, 2009, INMIC 2009, pp. 1–5. IEEE (2009)

8. Shafait, F., van Beusekom, J., Keysers, D., Breuel, T.M.: Document cleanup using page frame detection. Int. J. Doc. Anal. Recogn. (IJDAR) 11(2), 81–96 (2008)

9. Stamatopoulos, N., Gatos, B., Georgiou, T.: Page frame detection for double page document images. In: Proceedings of the 9th IAPR International Workshop on Document Analysis Systems, pp. 401–408. ACM (2010)

10. Stamatopoulos, N., Gatos, B., Kesidis, A.: Automatic borders detection of camera document images. In: 2nd International Workshop on Camera-Based Document Analysis and Recognition, Curitiba, Brazil, pp. 71–78 (2007)

11. Otsu, N.: A threshold selection method from gray-level histograms. Automatica 11(285–296), 23–27 (1975)

12. Canny, J.: A computational approach to edge detection. IEEE Trans. Pattern Anal. Mach. Intell. 8(6), 679–698 (1986)

13. Su, B., Lu, S., Tan, C.L.: Binarization of historical document images using the local maximum and minimum. In: Proceedings of the 9th IAPR International Workshop on Document Analysis Systems, pp. 159–166. ACM (2010)

14. Su, B., Lu, S., Tan, C.L.: Combination of document image binarization techniques. In: 2011 International Conference on Document Analysis and Recognition (ICDAR), pp. 22–26. IEEE (2011)

15. Gonzalez, R.C., Woods, R.E., Eddins, S.L.: Digital Image processing Using MATLAB. Gatesmark Publishing, Knoxville (2009)

Robust Binarization of Stereo and Monocular Document Images Using Percentile Filter

Muhammad Zeshan Afzal[1]($\boxtimes$), Martin Krämer[1], Syed Saqib Bukhari[1], Mohammad Reza Yousefi[1], Faisal Shafait[2], and Thomas M. Breuel[1]

[1] Technical University of Kaiserslautern, Kaiserslautern, Germany
{afzal,kraemer,bukhari,yousefi,tmb}@iupr.com
[2] The University of Western Australia, Crawley, Australia
faisal.shafait@uwa.edu.au

Abstract. Camera captured documents can be a difficult case for standard binarization algorithms. These algorithms are specifically tailored to the requirements of scanned documents which in general have uniform illumination and high resolution with negligible geometric artifacts. Contrary to this, camera captured images generally are low resolution, contain non-uniform illumination and also posses geometric artifacts. The most important artifact is the defocused or blurred text which is the result of the limited depth of field of the general purpose hand-held capturing devices. These artifacts could be reduced with controlled capture with a single camera but it is inevitable for the case of stereo document images even with the orthoparallel camera setup.

Existing methods for binarization require tuning for the parameters separately both for the left and the right images of a stereo pair. In this paper, an approach for binarization based on the local adaptive background estimation using percentile filter has been presented. The presented approach works reasonably well under the same set of parameters for both left and right images. It also shows competitive results for monocular images in comparison with standard binarization methods.

1 Introduction

The extensiveaut]Afzal Muhammad Zeshanaut]Krämer Martinaut]Bukhari Syed Saqibaut]Yousefi Mohammad Rezaaut]Shafait Faisalaut]Breuel Thomas M. use of portable cameras for capturing documents is driving the current research in the area. It is due to the inexpensiveness and ease of use of such devices. Although camera captured documents offer many advantages, they also have inherited problems because of capturing procedure and the devices themselves, which are used for capturing. The document image processing pipeline both for the monocular and stereo images [1–3] in most of the cases starts with the binarization of the document images in order to extract bi-level features for further processing.

Off-the-shelf passive sensing devices, e.g. customer grade hand-held cameras, can only focus objects which are at a certain distance from the camera. This is

M. Iwamura and F. Shafait (Eds.): CBDAR 2013, LNCS 8357, pp. 139–149, 2014.
DOI: 10.1007/978-3-319-05167-3_11, © Springer International Publishing Switzerland 2014

known as depth of field. The objects or the parts of objects which are nearer or farther from that end up being not properly focused in the captured image. Although it is possible to have a setup with large enough depth of field, it is not possible with customer consumer grade cameras, because it requires the knowledge about the actual scene properties, e.g. distance to the objects, and the camera properties, e.g. lens aperture etc. Another way to tackle this problem is to correct depth of field after acquiring the images, but it would require multiple captures depending upon the scene.

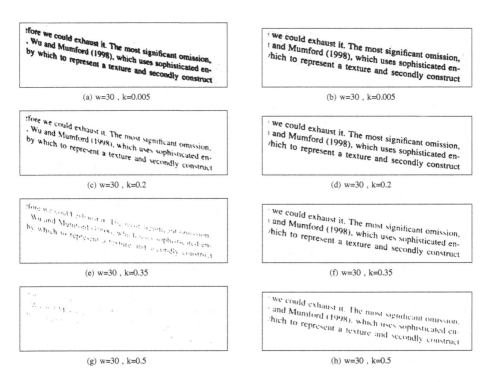

Fig. 1. The left column of the figure comprising of (a, c, e, g) shows the left image of the stereo pair with Savoula binarization evaluated using different k values with a fixed window size of 30. The right column of the figure comprising of (b, d, f, h) shows the right image of the stereo pair with Savoula binarization evaluated using different k values with fixed window size of 30.

We show the effect of binarization for the stereo pairs using local adaptive binarization approach Savoula as it has been reported to perform relatively better than other local approaches [4].

We consider a stereo image pair where the effect of blurring in the left image is major. In Fig. 1 the left column, i.e. Fig. 1(a, c, e and g) correspond to the left image of the stereo pair. In Fig. 1 the right column, i.e. Fig. 1(b, d, f and h) correspond to the right image of the stereo pair. The left image is blurred and the right image is correctly focused. The binarization for both of the images has been carried out using Savoula with a window size of 30 as it has been reported by Bukhari et al. [5] as suitable window size. The values of k are varied for both of the images. As we lower the value of k, which is depicted in Fig. 1a and b, the results of binarization become noisy. It contains salt and pepper noise. On the other hand increasing the value of k acts differently on left and right image. While the binarization of the right image remains reasonable under the changing the value of k, the left image, which is blurred, produces degraded binarization results and the foreground is vanishes with the increasing value of k.

This paper proposes an approach based on background estimation using percentile filters which performs reasonable binarization for both left and right images of stereo pair under the same set of parameters. The rest of this paper is organized as follows: the next section describes the related work for the approaches for binarization, the percentile filter and the stereo approaches document image processing which uses binarization. Section 3 describes the percentile filter and the proposed binarization algorithm. Section 4 shows the experimental results with quantitative evaluation and the paper ends with the conclusion presented in Sect. 5.

2   Related Work

The diversity of document images have been driving the research in document image processing in various directions. The researchers are trying to come up with generalized methods to be able to process a wide variety of documents. The binarization methods have also been proposed keeping in mind certain types of document images. The binarization of documents is aimed at either color [6–9] or gray [4,5,10–13] level images, which leads to the different methods of binarization. In general, when the image is being thresholded it could be done by determining a global threshold for whole of the page known as global binarization methods. On the other hand, the threshold can also be determined using only the statistics determined by a local window centered around the pixel which is being thresholded. A detailed discussion about the advantages and disadvantages of both local and global approaches is discussed in Bukhari et al. [5] which concludes that global binarization approaches, e.g. Otsu [10] shows a suboptimal performance for camera captured documents. It is due to the fact that there are certain variations, which appear only in specific parts of the image, e.g. a page might contain a defocused region and another could be illuminated differently, whereas other areas might have other geometric distortions. In contrast local methods can adapt themselves, depending upon the image characteristic of the local region. These methods could further be divided into two categories. The first category of methods is pixel-based and the other category is content-based [5]. In pixel based methods text and non-text regions are treated equally for determining the threshold used for local binarization. In content based methods the text and non-text regions are identified and different thresholds are used for text and non-text regions. The threshold in such methods is fixed for all the text regions. These method improve the performance, but the threshold is not adapted in accordance with background properties.

The proposed method in this paper takes into consideration the background statistics based on percentile filters [14]. So, this method can be categorized as a pixel based binarization method according to earlier classification. One very widely used example of percentile filter is median filter [14] which equals fifty (50) percentile. The percentile filters are generally categorized as ordered rank filters [15]. Rank order filter are very general and can be used to approximate other filters, e.g. median filter, as mentioned above, or morphological filters [16]. There are many ways how a percentile filter can be calculated. There are some methods which are specifically tailored for the images. One such comparison has been given by Duin et al. [17].

This paper describes work based on the percentile filter for binarization which works well both on focused and defocused images. This method also works well on monocular images with defocused parts.

3   Binarization Using Percentile Filter

Binarization using percentile filters starts with estimating the background at each location in the image. In a sense we are calculating a whole new image which is the background of the image based on percentile. First we define the percentile filter and after that the details and fast implementation are discussed and this section concludes with binarization details using percentile filters.

3.1   Percentile Filter

This algorithm has originally been proposed by [14]. We select a window of a certain size, defined by the user, and we calculate the histogram of the window. The window is defined as follows:

$$w(x,y) = (I_{ij})_{x-dx \leq i \leq x+dx, y-dy \leq j \leq y+dy} \tag{1}$$

where x and y denotes the location of the pixel at the center of the window and, dx and dy denote the size of the window both in x and y directions respectively.

Let us define the bounds of our window with the following sets

$$s_1 = \{x - dx \leq i \leq x + dx\} \tag{2}$$

$$s_2 = \{y - dy \leq j \leq y + dy\} \tag{3}$$

Now let us rewrite Eq. (1) with a single index.

$$w(x,y) = \{I_q \mid q = (i,j) \in s_1 \times s_2\} \tag{4}$$

To be able to calculate the percentile, let us sort the values in the window represented in Eq. (4) by defining an ordering function:

$$ord(a, b) = \begin{cases} 1, & \text{if } a > b \\ -1, & \text{else} \end{cases} \tag{5}$$

Let the number of pixels in the window be n and we define the following sequence:

$$ws(x, y) = (I_k)_{0 \leq k < n} \text{ such that } \forall k : ord(I_k, I_{k+1}) < 0 \tag{6}$$

The index of the percentile is given as follows:

$$i_p = p \times n/100 \text{ where } (0 \leq p \leq 100) \tag{7}$$

where i_p denotes the index of the value selected as a percentile, p is the required percentile and n is denoting the total number of elements.

Combining Eqs. (6) and (7) we define the value percentile for our window

$$ws_p(x, y) = I_{i_p} \tag{8}$$

where ws_p denotes the value of the pth percentile of the window centered around (x, y). It is important to note that all the things have been shown above for one window centered at (x, y). This procedure will be repeated for whole of the image. For determining the percentile of the pixels near the boundary, reflecting boundary conditions are used, i.e. the image has been mirrored to handle indices lying outside the image. An efficient implementation of the percentile filter based on histograms has been discussed in [17].

3.2 Binarization

A simple method of binarization using percentile filter is as follows: Let f be our original image and the domain of the image is all gray level values, i.e. $f(x, y) \in [0, 255]$ Let g be the background image estimated for each value based on percentile filters at every location (x, y) and the domain of the image corresponds to only two levels, i.e. $g(x, y) \in \{0, 255\}$. The background image is computed according to the procedure that has already been described. The thresholding has been done as follows:

$$o(x, y) = \begin{cases} 255, & \text{if } f(x, y) < t * g(x, y) \\ 0, & \text{otherwise} \end{cases} \tag{9}$$

where t is the parameter, which is used to determine that whether a pixel is foreground or background, depending on the similarity of the pixel, and the background, which has been estimated using percentile filter.

A more complicated version of the binarization procedure is described below step by step and is illustrated with Fig. 2. The image I is normalized in the range between 0 and 1 as follows

$$I_n = (I - I_{min})/I_{max} \tag{10}$$

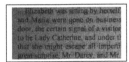

(a) Original image

(b) Normalized image

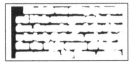

(c) After percentile filtering

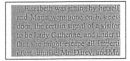

(d) Image after DOG

(e) Smoothed absolute magnitude

(f) Mask for test area detection

(g) After black and white clipping

(h) Final binarized image

Fig. 2. The steps of binarization are illustrated in the figures above. **(a)** Original image **(b)** Normalized version of the image shown in (a) according to the Eq. (10) **(c)** Resultant image after applying percentile filter to the normalized image shown in (b) according to the Eq. (11) **(d)** Resultant image after applying difference of Gaussian (DOG) to the image shown in (c) according to the Eq. (12) **(e)** The absolute values of the image shown in (d) with the smoothed with a Gaussian filter according to Eq. (13) **(f)** Binary mask depicting the text area from which the percentile values are calculated according to the Eq. (14) **(g)** Resultant image after applying the lo and hi percentile score to the image shown in (c) according to the Eq. (16) **(h)** Final image after the thresholding according to the Eq. (17)

A test image and its normalization have been shown in the Fig. 2a and b. We apply percentile filter with value of percentile p and window of size w

$$I_{p_w} = (P_{p_w}) * I_n \qquad (11)$$

Applying percentile filter on the normalized image shown in Fig. 2b results in the image shown in Fig. 2c. The value used of percentile filter is 80.

For selecting the low and high thresholds, the image is then enhanced using difference of Gaussian followed by a thresholding with a fixed value producing a binary image which is dilated. The first step is the image enhancement. Let σ_1 and σ_2 be the standard deviation of the successive levels which have been selected heuristically

$$I_g = (I_{p_w})_{\sigma_1} - (I_{p_w})_{\sigma_2} \qquad (12)$$

The result of applying difference of Gaussian to Fig. 2c is shown in Fig. 2d. The values 0 and 5 are used for the parameters σ_1 and σ_2 respectively.

The magnitude of the resultant image I_g is further smoothed with σ_2

$$I_s = (((I_g)^2)_{\sigma_2})^{0.5} \qquad (13)$$

The Fig. 2e shows the result of applying the above mentioned smoothing to the image shown in Fig. 2d.

For estimating the white and black clipping percentile a mask over the text area has been generated. For this purpose the resultant image from the step above is first thresholded with a constant value 0.3 to produced a binary image and the in the next step the binary image is dilated with a structuring element B

$$I_d = I_s \oplus B \tag{14}$$

The mask results in applying the above operations on Fig. 2e is shown in Fig. 2f. The length used for the structuring element is 10 both in the x and y directions. We use image I_d to mask I_{p_w} i.e

$$m(x, y) = \begin{cases} I_{p_w}(x, y), & \text{if } I_d(x, y) = 1 \\ 0, & \text{otherwise} \end{cases} \tag{15}$$

This means that the values, which are masked out, are not used for the calculation of the percentile score.

Let p_b and p_w be the black and white clipping percentile which are heuristically selected. Let (lo) and (hi) be the percentile score calculated based on the black and white clipping percentile respectively calculated from the masked image. We use the calculated percentile scores on the image I_{p_w} as follows

$$I_f(x, y) = \frac{I_{p_w}(x, y) - lo}{hi - lo} \tag{16}$$

For calculating the percentile scores for our image shown in Fig. 2e we used he values 5 and 90 for p_b and p_w respectively. The resultant image is shown in Fig. 2g after applying it to the image shown in Fig. 2c.

The binary image is produced by thresholding as follows

$$I_b(x, y) = \begin{cases} 0, & \text{if } I_f(x, y) > t \\ 255, & \text{otherwise} \end{cases} \tag{17}$$

The resultant binary image is shown in the Fig. 2h for $t = 0.55$.

4   Experiments and Results

First we consider the stereo images for comparing binarization results. Figure 3 shows the result of the percentile filter in the first row (Fig. 3a and b) for the stereo image pair considered in Fig. 1. The second row contains the result for Sauvola binarization of the same image pair. Figures in the second row, i.e. (Fig. 3c and d) are the same as (Fig. 1c and d) and shown here for comparison purposes. The results show that the percentile filter for both images perform better than Sauvola because the binarization is almost the same. This is essential for stereo matching. The Sauvola binarization performs well for the focused image as can be seen in Fig. 3d, but for the defocused image the quality is degraded and it might not help stereo matching for finding reliable matches.

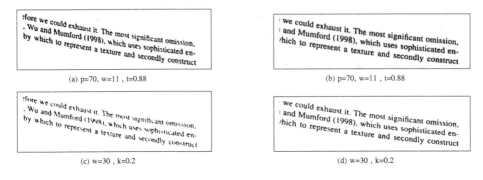

(a) p=70, w=11 , t=0.88   (b) p=70, w=11 , t=0.88

(c) w=30 , k=0.2   (d) w=30 , k=0.2

Fig. 3. The upper row (a, b) shows the left and right image of the stereo pair binarized using percentile filter with the same set of parameters, i.e. (p = 70, w = 11 , t = 0.85). The lower row (c, d) shows the left and right image of the stereo pair binarized using Savoula with the same set of parameters, i.e. (w = 30, k = 0.2). The percentile filter performs better on both left and the right images of the stereo pair with same set of parameters.

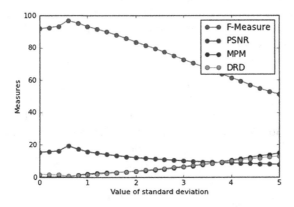

Fig. 4. Measures for the blurred image

In order to observe the effect of percentile filter for the blurred image restoration, we took a monocular image and its ground truth. The image has been convolved with an isotropic Gaussian for several values of standard deviation ranging from 0 to 5. Then the image is binarized using the percentile filter and the results are shown in Fig. 5 for the visual inspection. The full effect of the measures [18] has been shown in the Fig. 4. It can be observed that the percentile filter is robust against the blurring effect which could either be caused by the stereo cameras or in general by a single camera.

Furthermore, the proposed method has also been evaluated on monocular document images. The complicated version of binarization which helps us preserving the character shapes better. The dataset consists of 25 degraded images each of size 2000 × 500. We compare the results of our approach with standard Savoula binarization. For finding the best parameters for both of the methods

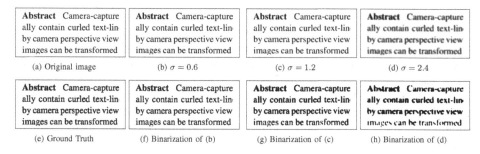

| (a) Original image | (b) $\sigma = 0.6$ | (c) $\sigma = 1.2$ | (d) $\sigma = 2.4$ |

| (e) Ground Truth | (f) Binarization of (b) | (g) Binarization of (c) | (h) Binarization of (d) |

Fig. 5. The upper row (a, b, c, d) shows the original image and the blurred image with standard deviation of 0.6, 1.2 and 1.4 respectively. The lower row (e, f, g, h) shows the ground truth and the restored images corresponding to their smoothed counter parts. The values of the parameters used for the binarization are (p = 33, w = 75, t = 0.78).

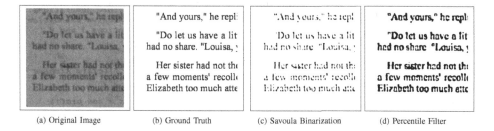

| (a) Original Image | (b) Ground Truth | (c) Savoula Binarization | (d) Percentile Filter |

Fig. 6. The comparison of the Savoula with percentile filter. The characters shapes are better using percentile filter in comparison to Savoula.

Table 1. Comparison of proposed method with Savoula Binarization. While Savoula performs a bit better for FMeasure evaluation, our method performs far better for OCR-Based evaluation

Measure	Savoula	Percentile
FMeasure	92.07	89.39
OCR error (edit distance)	69.29	31.40

we used FMeasure for error evaluation. The method of Savoula binarization has two parameters k and w. A grid search over an interval $[0.1, 0.35]$ with step size of 0.1 and $[3, 53]$ with step size of 2 respectively for k and w has been performed. The best values are 0.17 and 15 for k and w respectively. As we can see in the Table 1 that the performance of both methods for FMeasure is comparable and in fact Savoula performs a bit better than our method. On the other side, for OCR error measure, our method performs much better than Savoula because it is able to preserve the character shapes better which is very useful for practical applications. We used ocropus [19] for the OCR error measures. A sample binarized image is shown in Fig. 6.

5 Conclusion

A simple local binarization method is presented. We have shown applicability of the proposed binarization method on stereo document images. Compared to conventional binarization approaches, the main benefit is that the same parameters can be used for both images of the stereo image pair and still produce good binarization results. We have also shown that performance is comparable to standard methods for Fmeasure and our methods outperforms standard Savoula method by a big margin for OCR-Based evaluation.

References

1. Afzal, M., Krämer, M., Bukhari, S., Shafait, F., Breuel, T.: Improvements to uncalibrated feature-based stereo matching for document images by using text-line segmentation. In: Proceedings of the 10th IAPR International Workshop on Document Analysis Systems (2012)
2. Afzal, M., Bukhari, S., Krämer, M., Shafait, F., Breuel, T.: Robust stereo matching for document images using parameter selection of text-line extraction. In: 21st International Conference on Pattern Recognition, ICPR'12, Tsukuba, Japan, November 2012
3. Krämer, M., Afzal, M., Bukhari, S., Shafait, F., Breuel, T.: Robust stereo correspondence for documents by matching connected components of text-lines with dynamic programming. In: 21st International Conference on Pattern Recognition, ICPR'12, Tsukuba, Japan, November 2012
4. Sauvola, J., Pietikäinen, M.: Adaptive document image binarization. Pattern Recogn. **33**, 225–236 (2000)
5. Bukhari, S.S., Shafait, F., Breuel, T.: Adaptive binarization of unconstrained handheld camera-captured document images. J. Univ. Comput. Sci. **15**(18), 3343–3363 (2009)
6. Sobottka, K., Kronenberg, H., Perroud, T., Bunke, H.: Text extraction from colored book and journal covers. IJDAR **2**(4), 163–176 (2000)
7. Tsai, C.-M., Lee, H.-J.: Binarization of color document images via luminance and saturation color features. IEEE Trans. Image Process. **11**(4), 434–451 (2002)
8. Badekas, E., Nikolaou, N.A., Papamarkos, N.: Text localization and binarization in complex color documents. In: MLDM Posters, pp. 1–15 (2007)
9. Orii, H., Kawano, H., Maeda, H., Ikoma, N.: Text-color-independent binarization for degraded document image based on map-mrf approach. IEICE Trans. **94–A**(11), 2342–2349 (2011)
10. Otsu, N.: A threshold selection method from gray-level histograms. IEEE Trans. Syst. Man Cybern. **9**(1), 62–66 (1979)
11. Gatos, B., Pratikakis, I., Perantonis, S.J.: Adaptive degraded document image binarization. Pattern Recogn. **39**(3), 317–327 (2006)
12. Shafait, F., Keysers, D., Breuel, T.: Efficient implementation of local adaptive thresholding techniques using integral images. In: Proceedings of the 15th Document Recognition and Retrieval Conference, Part of the IST/SPIE International Symposium on Electronic Imaging, January 26–31, San Jose, CA, USA, vol. 6815. SPIE, January 2008

13. Rivest-Hénault, D., Moghaddam, R.F., Cheriet, M.: A local linear level set method for the binarization of degraded historical document images. IJDAR **15**(2), 101–124 (2012)
14. Justusson, B.: Median filtering: statistical properties. In: Two-Dimensional Digital Signal Prcessing II. Topics in Applied Physics, pp. 161–196. Springer, Heidelberg (1981)
15. Heygster, G.:
16. Soille, P.: On morphological operators based on rank filters. Pattern Recogn. **35**(2), 527–535 (2002)
17. Duin, R., Haringa, H., Zeelen, R.: Fast percentile filtering. Pattern Recogn. Lett. **4**(4), 269–272 (1986)
18. Pratikakis, I., Gatos, B., Ntirogiannis, K.: Icdar 2011 document image binarization contest (dibco 2011). In: 2011 International Conference on Document Analysis and Recognition (ICDAR), pp. 1506–1510, September 2011
19. Breuel, T.M.: The OCRopus Open Source OCR System. http://code.google.com/p/ocropus/

Hyperspectral Document Imaging: Challenges and Perspectives

Zohaib Khan$^{(\boxtimes)}$, Faisal Shafait, and Ajmal Mian

School of Computer Science and Software Engineering,
The University of Western Australia,
35 Stirling Highway, Crawley, WA 6009, Australia
zohaib@csse.uwa.edu.au

Abstract. Hyperspectral imaging provides measurement of a scene in contiguous bands across the electromagnetic spectrum. It is an effective sensing technology having vast applications in agriculture, archeology, surveillance, medicine and forensics. Traditional document imaging has been centered around monochromatic or trichromatic (RGB) sensing often through a scanning device. Cameras have emerged in the last decade as an alternative to scanners for capturing document images. However, the focus has remained on mono-/tri-chromatic imaging. In this paper, we explore the new paradigm of hyperspectral imaging for document capture. We outline and discuss the key components of a hyperspectral document imaging system, which offers new challenges and perspectives. We discuss the issues of filter transmittance and spatial/spectral non-uniformity of the illumination and propose possible solutions via pre and post processing. As a sample application, the proposed imaging system is applied to the task of writing ink mismatch detection in documents on a newly collected database (UWA Writing Ink Hyperspectral Image Database http://www.csse.uwa.edu.au/%7Eajmal/databases.html). The results demonstrate the strength of hyperspectral imaging in capturing minute differences in spectra of different inks that are very hard to distinguish using traditional RGB imaging.

Keywords: Hyperspectral document analysis · Forensic document examination · Ink mismatch detection

1 Introduction

Image scanning devices are currently the major source of creating digitized versions of documents both black and white as well as color. Traditional scanners are fairly limited with regards to the color information that they can capture as their imaging systems are designed to replicate the trichromatic RGB human visual system. In many situations high fidelity spectral information can be can be very useful, for example where it is required to distinguish between two similar inks [1] or determine the age of a writing or the document itself.

M. Iwamura and F. Shafait (Eds.): CBDAR 2013, LNCS 8357, pp. 150–163, 2014.
DOI: 10.1007/978-3-319-05167-3_12, © Springer International Publishing Switzerland 2014

Natural materials exhibit a characteristic spectral response to incident light. The spectral response of a material is responsible for its specific color. It is a signature property which can be used for material identification. Spectral imaging is an effective technique for measurement of the spectra of objects in the real world. A hyperspectral (HS) image of a scene is a series of contiguous narrowband images in the electro-magnetic spectrum. In contrast to a three channel RGB image, an HS image captures finer spectral information of a scene.

Satellite based multispectral imaging sensors have long been used for astronomical and remote sensing applications. Due to the high cost and complexity of these multispectral imaging sensors, various techniques have been proposed to utilize conventional imaging systems combined with a few off the shelf optical devices for multispectral imaging. In this paper, we discuss new challenges in the development of hyperspectral document imaging system. Various spectral imaging techniques have been developed over the years. An overview about different technologies for capturing hyperspectral images is given in Sect. 2. We focus on the HS imaging specific issues of spatial/spectral illumination variation and filter transmission variation and propose possible solutions to reduce these artifacts in Sect. 3. We apply the proposed HS imaging system to the task of ink mismatch detection (Sect. 4) on a newly developed writing ink hyperspectral image database. The paper is concluded in Sect. 5.

2   Overview of Hyperspectral Imaging

Strictly speaking, an RGB image is a three channel spectral image. An image acquired at more than three specific wavelengths in a band is referred to as a *Multispectral Image*. Generally, multispectral imaging sensors acquire more than three spectral bands. An image having finer spectral resolution or higher number of bands is regarded as a *Hyperspectral Image*. There is no clear demarcation with regards to the number of spectral bands/resolution between multispectral and hyperspectral images. However, hyperspectral sensors may acquire a few dozen to several hundred spectral measurements per scene point. For example, the AVIRIS (Airborne Visible/Infrared Imaging Spectrometer) of NASA has 224 bands in 400–2500 nm range [2].

A hyperspectral image has three dimensions: two spatial dimensions (x, y) and one spectral dimension (λ) as shown in Fig. 1. A hyperspectral image can be presented in the form of a *Hyperspectral Cube*. The basic concept for capturing hyperspectral images is to filter incoming light by the use of bandpass filters or dispersion optics. In the following we present a brief overview of different methods/technologies used for hyperspectral imaging, categorized based on the underlying optical phenomenon of bandpass filtering or chromatic dispersion. The overview presented here is limited to the hyperspectral imaging systems used in ground-based computer vision applications. Therefore, high cost and complex sensors for remote sensing employed in astronomy and other geo-spatial applications are not considered.

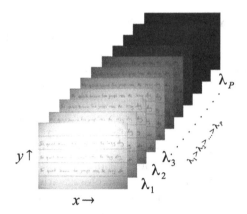

Fig. 1. A hyperspectral image is illustrated as a series of images along the spectral dimension.

2.1 Bandpass Filtering

In filter based approach, the objective is to allow light in a specific wavelength range to pass through the filter and reach the imaging sensor. This phenomenon is illustrated in Fig. 2. This can be achieved by using optical devices generally named bandpass filters or simply filters. The filters can be categorized into two types depending on the filter operating mechanism. The first type is the tunable filter or specifically the electrically tunable filter. The pass-band of such filters can be electronically tuned at a very high speed which allows for measurement of hyperspectral data in a wide range of wavelengths. The second type is the *non-tunable* filters. Such filters have a fixed pass-band of frequencies and are not recommended for use in time constrained applications. These filters require physical replacement either manually, or mechanically by a filter wheel. However, they are easy to use in relatively simple and unconstrained applications.

Non-Tunable Filters. A common approach to acquire multispectral images is by sequential replacement of bandpass filters between a scene and the imaging sensor. The process of filter replacement can be mechanized by using a wheel of filters. Such filters are useful where time factor is not involved and the goal is to image a static scene. Kise et al. [3] developed a three band multispectral imaging system by using interchangeable filter design; two in the visible range (400–700 nm) and one in the near infrared range (700–1000 nm). The interchangeable filters allowed for selection of three bands. The prototype was applied to the task of poultry contamination detection.

Tunable Filters. Electronically tunable filters come in different base technologies. One of the most common is the *Liquid Crystal Tunable Filter* (LCTF). The LCTF is characterized by its low cost, high throughput and slow tuning time. On the other hand, the *Acousto-Optical Tunable Filter* is known for high cost,

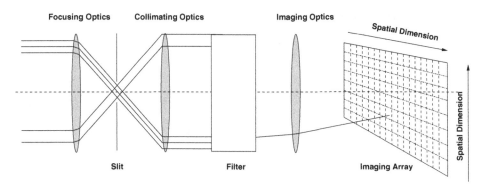

Fig. 2. In bandpass filtering, the filter allows only a specific wavelength of light to pass through, resulting in a single projection of the scene at a particular frequency.

low throughput, and faster tuning time. For a detailed description of the composition and operating principles of the tunable filters, the readers are encouraged to read [4,5].

Fiorentin et al. [6] developed a hyperspectral imaging system using a combination of CCD camera and LCTF in the visible range with a resolution of 5 nm. The device was used in the analysis of accelerated aging of printing color inks. The system is also applicable of monitoring the variation (especially fading) of color in artworks with the passage of time. The idea can be extended to other materials that may exhibit changes due to exposure to artificial or daylight illumination, such as document paper and ink.

Comelli et al. [7] developed a portable UV-fluorescence hyperspectral imaging system to analyze painted surfaces. The imaging setup comprises a UV-florescence source, an LCTF and a low noise CCD sensor. A total of 33 spectral images in the range (400–720 nm) in 10 nm steps were captured. The accuracy of the system was determined by comparison with the fluorescence spectra of three commercially available fluorescent samples measured with a bench-top spectro-fluorometer. The system was tested on a 15th century renaissance painting to reveal latent information related to the pigments used for finishing decorations in painting at various times.

2.2 Chromatic Dispersion

In dispersion based filtering, the objective is to decompose an incoming ray of light into its spectral constituent as shown in Fig. 3. This can be achieved by optical devices like diffraction gratings, prisms, *grisms* (grating and prism combined) and interferometers. We further outline chromatic dispersion based on refraction or interferometric optics.

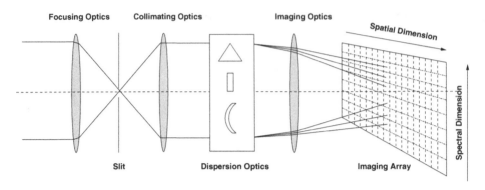

Fig. 3. In chromatic dispersion, the dispersion optics disperses the incoming light into its constituents which are projected onto the imaging plane.

Refraction Optics. Refraction is an intrinsic property of glass-like materials such as prisms. A prism separates the incoming light ray into its constituent colors. Du et al. [8] proposed a prism-based multispectral imaging system in the visible and infrared bands. The system used an occlusion mask, a triangular prism and a monochromatic camera to capture multispectral image of a scene. Multispectral images were captured at high spectral resolution while trading off the spatial resolution. The use of occlusion mask also reduced the amount of light available to the camera and thus decreased the signal to noise ratio (SNR). The prototype was evaluated for the tasks of human skin detection and physical material discrimination.

Gorman et al. [9] developed an *Image Replicating Imaging Spectrometer* (IRIS) using an arrangement of a *Birefringent Spectral De-multiplexer* (BSD) and off-the-shelf compound lenses to disperse the incoming light into its spectral components. The system was able to acquire spectral images in a snapshot. It could be configured to capture 8, 16 or 32 bands by increasing the number of stages of the BSD. It has, however, a Field-of-View limited by the width of a prism used in the BSD. A high spectral resolution is achieved by trading-off spatial resolution since a 2D detector is used.

Interferometric Optics. The optics such as interferometers can also be used as light dispersion devices by constructive and destructive interference. Burns et al. [10] developed a seven-channel multispectral imaging device using 50 nm bandwidth interference filters and a standard CCD camera. Mohan et al. proposed the idea of *Agile Spectral Imaging* [11]. Using a diffraction grating to disperse the incoming rays, a geometrical mask pattern was used to allow specific wavelengths to pass through and reach the sensor.

Descour et al. [12] presented a *Computed Tomography Imaging Spectrometer (CTIS)* design using three sinusoidal phase gratings to disperse light into multiple directions and diffraction orders. Assuming the dispersed images to be two dimensional projections of three dimensional multispectral cube, the multispectral cube is reconstructed using maximum-likelihood expectation maximization algorithm assuming Poisson likelihood law. The prototype works in the visible range (470–770 nm) and is able to reconstruct multispectral images of a simple target.

3   Hyperspectral Document Imaging

By carefully analyzing different technologies/methods for capturing hyperspectral images, we chose the tunable filter due to its easy integration with off-the-shelf machine vision cameras and programmatic control over the hardware (e.g. exposure time, spectral resolution, etc.). This section provides an overview of our hyperspectral document imaging setup and presents our approach for tackling various hyperspectral-imaging-specific challenges.

3.1   Acquisition Setup

Our system comprises of a monochrome machine vision CCD camera at a base resolution of 752×480 pixels. A focusing lens (1:1.4/16 mm) lies in front of the CCD camera. In order to capture images in discrete wavelength channels, a Liquid Crystal Tunable Filter (LCTF) is placed in front of the lens as shown in Fig. 4. The filter can tune to any wavelength in the visible range (400–720 nm) with an average tuning time of 50 ms. The bandwidth of the filter varies with the center wavelength, such that it is low at shorter wavelengths and high at longer wavelengths as shown in Fig. 5. It is measured in terms of *Full Width at Half Maximum (FWHM)* which ranges from 7 to 20 nm corresponding to 400 and 720 nm. Thus, the first few bands have very low SNR combined with the filter transmission loss (see Fig. 5). To compensate for the low SNR images, the document is illuminated by two halogen lamps.

To achieve sufficient fidelity in the spectral dimension, we capture hyperspectral images comprising 33 bands in the visible range (400–720 nm at steps of 10 nm). The target is captured in a sequential manner so that the total capture time is the sum of acquisition and filter tuning time for each band (5 s, several times faster than a commercial system [13]).

3.2   Compensation for Filter Transmittance

Typically, each band of a hyperspectral image is captured with a constant exposure time. Since different spectral bands are captured sequentially in our imaging setup, it is possible to vary exposure before each acquisition is triggered. Looking at the filter response at different wavelengths in Fig. 5, it can be observed that the amount of light transmitted is a function of the wavelength such that – with some minor glitches – the longer the wavelength λ, the higher the transmittance $\tau(\lambda)$. Extremely small values of $\tau(\lambda)$ for $\lambda \in [400, 450]$ result in insufficient

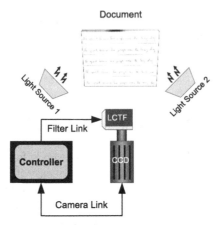

Fig. 4. An illustration of the proposed hyperspectral document image acquisition setup. The controller triggers cycles of filter tuning/image acquisition at a high speed allowing for efficient image capture.

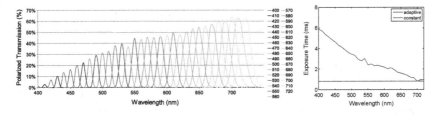

Fig. 5. Transmission functions of the LCTF at 10 nm wavelength step (left). Exposure time as a function of wavelength (right). Observe that the filter transmission at shorter wavelengths needs compensation.

energy captured by the imaging system in those bands corresponding to the blue region of the spectrum (see Fig. 1). To compensate for this effect, we model the exposure time $t_e(\lambda)$ as an inverse function of the wavelength such that the shorter the wavelength, the longer the exposure time:

$$t_e(\lambda) = \alpha(\tau_{\max} - \tau(\lambda)) + \bar{t}_e \qquad (1)$$

where $\tau_{\max}$ is the maximum transmission of the filter at any wavelength (i.e. transmission at $\lambda = 700$ nm for the filter used in this work – see Fig. 5), $\bar{t}_e$ is the corresponding exposure time, and α is a balancing coefficient. $\bar{t}_e$ is computed as the maximum possible exposure time for the band corresponding to $\tau_{\max}$ which ensures no image saturation. In order to keep each band unsaturated, we keep α to be small ($\alpha = 8$ in this work) and experimentally find a suitable value for $\bar{t}_e$.

3.3   Compensation for Non-Uniform Illumination Intensity

In hyperspectral document imaging, the use of a nearby illumination source induces a scalar field over the target image. This means that there is a spatially non-uniform variation in illumination. The result is that the pixels near the center of the image will be brighter (have higher energy) as compared to the pixels farther away towards the edges. This effect can be seen in Fig. 1. Let $\mathbf{p}(x, y)$ be the spectral response at the image location (x, y). It can be reasonably assumed here that the non-uniformity in illumination is only a function of pixel coordinates (x, y) and does not depend on the wavelength λ. This assumption will hold for each (x, y) as long as $\mathbf{p}(x, y)$ is not saturated. Hence, normalizing the spectral response at each pixel to the unit vector:

$$\hat{\mathbf{p}}(x, y) = \frac{\mathbf{p}(x, y)}{\|\mathbf{p}(x, y)\|} \tag{2}$$

will largely compensate for the effect of non-uniform illumination intensity.

3.4   Compensation for Illuminant's Non-Uniform Spectral Power Distribution

Assuming Lambertian surface reflectance, the hyperspectral image of a document can be modeled as follows. The formation of an N channel hyperspectral image $\mathcal{I}(x, y, \lambda), \lambda = 1, 2, ..., N$ of a document is mainly dependent on four factors: the illuminant spectral power distribution $\mathcal{L}(\lambda)$, the scene spectral reflectance $\mathcal{S}(x, y, \lambda)$, the filter transmittance $\tau(\lambda)$, and the sensor spectral sensitivity $\mathcal{C}(\lambda)$. Hence, image intensity of a particular spectral band λ can be calculated as

$$\mathcal{I}(x, y, \lambda) = \int_{\lambda_{\min}}^{\lambda_{\max}} \mathcal{L}(\lambda)\mathcal{S}(x, y, \lambda)\tau(\lambda)\mathcal{C}(\lambda)d\lambda \tag{3}$$

where $\lambda_{\min}$ and $\lambda_{\max}$ define the bandwidth of the spectral band λ.

Most of the illumination sources do not have a flat power distribution across different wavelengths (see Fig. 6 for spectral power distribution of some common illuminants). To compensate for non-uniform spectral power distribution of the illuminant, color constancy methods are applied. Van de Weijer et al. [14] proposed a unified formulation for different color constancy algorithms. Varying the parameters of the following formulation, leads to estimation of the illuminant spectra

$$\hat{\mathcal{L}}(\lambda : n, p, \sigma) = \frac{1}{\kappa} \left(\int_y \int_x |\nabla^n \mathcal{I}_\sigma(x, y)|^p dx \, dy \right), \tag{4}$$

where n is the order of differential, p is the Minkowski norm and σ is the scale of the Gaussian filter. $\mathcal{I}_\sigma(x, y) = \mathcal{I}(x, y) * G(x, y : \sigma)$ is the Gaussian filtered image. κ is a constant, chosen such that the estimated illuminant spectra has a unit ℓ_2-norm. The illumination corrected hyperspectral image is obtained by a simplified linear transformation

$$\hat{\mathcal{I}}(x, y, \lambda) = \mathcal{M}\mathcal{I}(x, y, \lambda), \quad \mathcal{M} \in \mathbb{R}^{N \times N}, \tag{5}$$

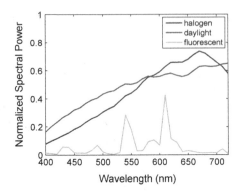

Fig. 6. Spectral power distributions of various illuminant sources. Observe that the low illuminant power at shorter wavelengths needs compensation.

where $\mathcal{M}$ is a diagonal matrix such that

$$\mathcal{M}_{i,j} = \begin{cases} 1/\mathcal{L}(\lambda_i) & \text{if } i = j \\ 0 & \text{otherwise} \end{cases} \tag{6}$$

Color constancy can be achieved by making assumptions on the first or higher order statistics of the image. There is no strict rule as to which assumption is the best. Rather it mainly depends which particular assumption suits the given image content. Following is a brief overview of assumptions made by different color constancy algorithms.

Gray World (GW) algorithm [15] assumes that the average image spectra is gray, so that the illuminant spectra can be estimated as the deviation from the gray of average.

Gray Edge (GE) algorithm [14] assumes that the mean spectra of the edges is gray so that the illuminant spectra can be estimated as the shift from gray of the mean of the edges.

White Point algorithm [16] assumes the presence of a white patch in the scene such that the maximum value in each channel is the reflection of the illuminant from that white patch.

Shades-of-Gray (SoG) algorithm [17] is based on the assumption that the ℓ_p-norm of a scene is a shade of gray.

general Gray World (gGW) algorithm [15] is based on the assumption that the ℓ_p-norm of a scene after smoothing is gray.

Based on the assumptions behind each of these algorithms, the *White Point* algorithm appears to be the most appropriate for estimating illuminant spectral power distribution from document images. Since documents are often printed on white paper, the assumption made by the WP algorithm about the presence of a white patch in the image would be mostly satisfied.

4   Application to Ink Mismatch Detection

As a sample application of hyperspectral imaging in document analysis, we chose ink mismatch detection (please refer to [1] for more details). In this paper, we specifically address the challenges associated with hyperspectral imaging of documents. The main focus is on compensating the effects of spatial and spectral non-uniformity of illumination. We perform additional experiments to observe the effects of illumination normalization using proposed compensation techniques. Using the imaging setup described in Sect. 3.1, a database consisting of 70 hyperspectral images of a hand-written note in 10 different inks by 7 subjects was collected. All subjects were instructed to write the same sentence, once in each ink on a white paper. The pens included 5 varieties of blue ink and 5 varieties of blank ink pens. It was ensured that the pens came from different manufacturers while the inks still appeared visually similar. Then, we produced mixed writing ink images from single ink notes by joining equally sized image portions from two inks written by the same subject. This made roughly the same proportion of the two inks under question.

The pre-processed mixed-ink images were first binarized using an adaptive thresholding method [18] and then fed to the k-means clustering algorithm with a fixed value of $k = 2$. Finally, ink mismatch detection accuracy was computed as

$$\text{Accuracy} = \frac{\text{True Positives}}{\text{True Positives} + \text{False Positives} + \text{False Negatives}}$$

The mismatch detection accuracy is averaged over seven samples for each ink combination C_{ij}. It is important to note that according to this evaluation metric, the accuracy of a random guess (in a two class problem) will be $1/3$. This is different to common classification accuracy metrics where the accuracy of a random guess is $1/2$. This is because our chosen metric additionally penalizes false negatives which is critical to observe in a our problem.

As discussed in Sect. 3.3, a spatially varying illumination is not desirable and modulates the spectral responses of the image pixels. In order to undo the effect of a non-uniform illumination, the images are normalized using Eq. 2. Figure 7 presents the mismatch detection accuracies on raw and normalized hyperspectral images. The improvement in correctly segmenting mismatching inks is highly evident for a majority of ink combinations of the blue and black ink, respectively.

In Sect. 3.2, an adaptive exposure scheme was proposed to compensate for the varying filter transmittance. The adaptive exposure results in a higher SNR for bands with a low transmittance. We compare the use of adaptive exposure with constant exposure for hyperspectral ink mismatch detection. It can be noticed from Fig. 8 that the use of adaptive exposure either slightly improves the accuracy or remains close to the performance achieved by constant exposure.

We now evaluate the ink mismatch detection accuracy after compensating for illuminant spectral non-uniformity (color constancy) to that of no compensation as outlined in Sect. 3.4. It can be seen in Fig. 9 that there is only a slight improvement for some of the ink combinations after using color constancy.

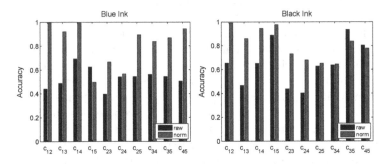

Fig. 7. Comparison of ink mismatch detection accuracies between raw and normalized (using Eq. 2) images. Note that the normalization significantly improves accuracy.

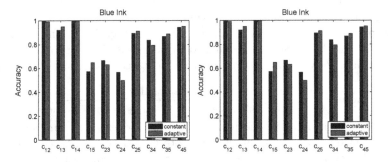

Fig. 8. Comparison of mismatch detection accuracies using constant or adaptive (using Eq. 1) exposure. Observe that the adaptive exposure strategy results in a more accurate discrimination between inks of the same color.

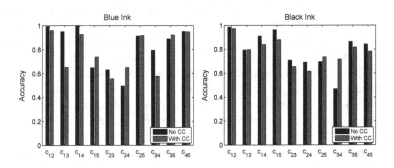

Fig. 9. Comparison of ink mismatch detection accuracies between *no cc* and *cc* (using Eq. 5) images. Note that the normalization by color constancy does not improve mismatch detection accuracy.

The efficacy of the proposed hyperspectral document imaging system can be visually appreciated by a qualitative analysis of the example images. Figure 10 shows two example images of blue and black inks. The images are made by joining samples of ink 1 and ink 2 for both blue and black inks, separately. The original images are shown in RGB for clarity. The ground truth images are labeled in different colors to identify the constituent inks in the mixture.

Observe that the raw HS images are yellowish due to the strong illuminant bias as well as low filter transmittance for the wavelengths in the blue spectrum range. Besides, spatial non-uniformity of the illumination can be observed from the center to the edges. The mismatch detection results on raw images indicate that the clustering is biased by the illumination intensity, instead of the ink color. After normalization of the raw HS images, it is evident that the effect of illumination is highly depreciated. This results in an accurate mismatch detection result that closely follows the ground truth.

We finally observe the effect of color constancy on ink mismatch detection. Notice that the mismatch detection result is largely unaffected except for a few noisy pixels which are misidentified as being from a different ink. One of the clear benefits of color constancy is that it highly improves the visual appearance of the images by removing the illumination bias.

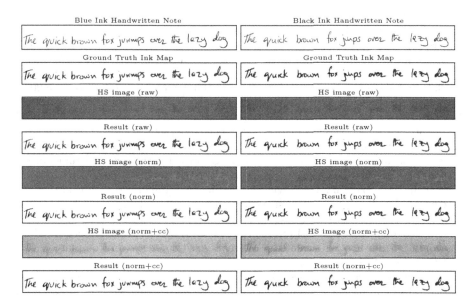

Fig. 10. An illustration of ink mismatch detection on a blue ink and a black ink handwritten notes, acquired using adaptive exposure. The ground truth ink pixels are labeled in pseudo colors (red: ink 1, green: ink 2). The spatially non-uniform illumination pattern can be observed in raw HS images, with high energy in the center and low towards the edges. Normalization removes the illumination bias and greatly improves segmentation accuracy. Color constancy improves the visualization of HS images, while resulting in comparable accuracy.

5   Conclusion and Future Work

Hyperspectral imaging of documents has potentially numerous applications in document analysis. The spatial non-uniformity of illuminant source was compensated to a great extent by the proposed normalization strategy. The variable filter transmission was compensated for by a linear adaptive exposure function. Further improvements could be expected by introducing non-linear adaptive exposure functions. We also explored color constancy for illuminant spectral normalization which greatly improved the HS image visualization. More research attention is required to the limitations of current hardware to address challenges of illumination variation and variable filter transmission.

Acknowledgment. This research work was partially funded by the ARC Grant DP110102399 and the UWA Grant 00609 10300067.

References

1. Khan, Z., Shafait, F., Mian, A.: Hyperspectral imaging for ink mismatch detection. In: Proceedings of the International Conference on Document Analysis and Recognition (ICDAR) (2013)
2. Shippert, P.: Introduction to hyperspectral image analysis. Online J. Space Commun. **3**, 1–13 (2003)
3. Kise, M., Park, B., Heitschmidt, G.W., Lawrence, K.C., Windham, W.R.: Multispectral imaging system with interchangeable filter design. Comput. Electron. Agric. **72**(2), 61–68 (2010)
4. Gat, N.: Imaging spectroscopy using tunable filters: a review. In: AeroSense 2000, International Society for Optics and Photonics, pp. 50–64 (2000)
5. Poger, S., Angelopoulou, E.: Multispectral sensors in computer vision. Technical Report CS-2001-3, Stevens Institute of Technology (2001)
6. Fiorentin, P., Pedrotti, E., Scroccaro, A.: A multispectral imaging device for monitoring of colour in art works. In: Proceedings of the International Instrumentation and Measurement Technology Conference (I2MTC), pp. 356–360. IEEE (2009)
7. Comelli, D., Valentini, G., Nevin, A., Farina, A., Toniolo, L., Cubeddu, R.: A portable UV-fluorescence multispectral imaging system for the analysis of painted surfaces. Rev. Sci. Instrum. **79**(8), 086112 (2008)
8. Du, H., Tong, X., Cao, X., Lin, S.: A prism-based system for multispectral video acquisition. In: Proceedings of the International Conference on Computer Vision (ICCV), pp. 175–182 (2009)
9. Gorman, A., Fletcher-Holmes, D.W., Harvey, A.R., et al.: Generalization of the Lyot filter and its application to snapshot spectral imaging. Opt. Express **18**(6), 5602–5608 (2010)
10. Burns, P.D., Berns, R.S.: Analysis of multispectral image capture. In: Proceedings of the 4th IS&T/SID Color Imaging Conference, pp. 19–22 (1996)
11. Mohan, A., Raskar, R., Tumblin, J.: Agile spectrum imaging: programmable wavelength modulation for cameras and projectors. Comput. Graph. Forum **27**(2), 709–717 (2008)
12. Descour, M., Dereniak, E.: Computed-tomography imaging spectrometer: experimental calibration and reconstruction results. Appl. Opt. **34**(22), 4817–4826 (1995)

13. Joo Kim, S., Deng, F., Brown, M.S.: Visual enhancement of old documents with hyperspectral imaging. Pattern Recogn. **44**(7), 1461–1469 (2011)
14. Van De Weijer, J., Gevers, T., Gijsenij, A.: Edge-based color constancy. IEEE Trans. Image Process. **16**(9), 2207–2214 (2007)
15. Buchsbaum, G.: A spatial processor model for object colour perception. J. Franklin inst. **310**(1), 1–26 (1980)
16. Land, E.: The retinex theory of color vision. Science Center, Harvard University (1974)
17. Finlayson, G., Trezzi, E.: Shades of gray and colour constancy. In: Twelfth Color Imaging Conference: Color Science and Engineering Systems, Technologies, and Applications, pp. 37–41 (2004)
18. Shafait, F., Keysers, D., Breuel, T.M.: Efficient implementation of local adaptive thresholding techniques using integral images. In: Document Recognition and Retrieval XV, pp. 681510–681510-6 (2008)

Mobile Phone Camera-Based Video Scanning of Paper Documents

Muhammad Muzzamil Luqman$^{(\boxtimes)}$, Petra Gomez-Krämer,
and Jean-Marc Ogier

L3i Laboratory, University of La Rochelle, Avenue M. Crépeau,
17042 La Rochelle, France
{muhammad_muzzamil.luqman,petra.gomez,jean-marc.ogier}@univ-lr.fr

Abstract. Mobile phone camera-based document video scanning is an interesting research problem which has entered into a new era with the emergence of widely used, processing capable and motion sensors equipped smartphones. We present our ongoing research on mobile phone camera-based document image mosaic reconstruction method for video scanning of paper documents. In this work, we have optimized the classic keypoint feature descriptor-based image registration method, by employing the accelerometer and gyroscope sensor data. Experimental results are evaluated using optical character recognition (OCR) on the reconstructed mosaic from mobile phone camera-based video scanning of paper documents.

Keywords: Camera-based document image analysis · Document image mosaicing · Image registration

1 Introduction

In recent years, the availability of camera equipped, processing capable, inertial sensors fitted and moderate priced mobile phones (a.k.a. smartphones), has attracted the attention of the research community to employ them for complementing the classical document scanning devices. The use of these devices for document scanning provides interesting advantages over the traditional document scanning devices. They can be used to scan thick books, historical documents that are too fragile to touch, text in scenes (walls, whiteboards, etc.), and large sized documents [8]. However, the use of smartphones introduces new challenges to document scanning which are not faced by classical document scanning devices. These challenges include uneven lighting, perspective distortion, non-planer surfaces, motion blur and low resolution of the cameras [4,8].

In this paper we present our ongoing work on mobile phone camera-based video scanning of paper documents. The video scanning of a paper document is achieved by swiping the mobile phone camera over the paper document and recording the accelerometer sensor data along with capturing the video frames. During the video scanning the orientation of the mobile phone camera is obtained

M. Iwamura and F. Shafait (Eds.): CBDAR 2013, LNCS 8357, pp. 164–178, 2014.
DOI: 10.1007/978-3-319-05167-3_13, © Springer International Publishing Switzerland 2014

from the gyroscope data and the user is provided with visual feedback on the orientation of the phone to avoid perspective distortion. A complete mosaic image of the paper document is reconstructed from the captured video frames by employing an optimized keypoint feature descriptor-based image registration technique. The optimization is achieved by employing the recorded accelerometer sensor data. The resulting reconstructed mosaic has a higher resolution than a simple photo of the document taken by the same camera.

In literature the camera-based scanning of paper documents has been approached by various works which are mainly motivated by panorama reconstruction and image mosaicing techniques from the computer vision research community. In [4] first an image feature-based technique is used to estimate the camera motion and to assist the user to capture images of patches of document. The estimated camera motion is used with a keypoint feature descriptor-based technique for registration of captured image patches and reconstruction of a mosaic of the document. In [14] an algorithm for 2D scanning of a planar scene is proposed. The topology of the video frames are inferred on a 2D manifold by alignment of successive video frames and overlapping video frames. The aligned frames are merged by using a multi-resolution method for constructing a seamless mosaic. In [11] local likely arrangement hashing (LLAH), which is originally an image retrieval technique, is used for keypoint detection and feature description in frames. Images are aligned by matching LLAH feature descriptors and the feature correspondences are used for combining input frames for reconstruction of mosaic. In [7], first, captured frames are rectified by removing perspective distortion using texture flow information. A Hough transform-based voting scheme is used for finding translation and scaling between video frames. The reconstruction of the mosaic is achieved by a sharpness-based seamless composition of overlapping images. In [17] inertial sensors in mobile phones have been employed for constructing panoramas on mobile phones. In a first step the position and relative displacement of video frames are computed by inertial sensor data. Using the alignment estimation from inertial sensor data, a more precise alignment of the video frames is computed by using a keypoint feature descriptor-based technique and the mosaic image is constructed by using the feature correspondences.

The perspective distortion is very important to be handled in case of camera-based document scanning. A document image mosaicing technique should directly or indirectly rectify perspective distortion of the captured frames before reconstructing the mosaic image. A summary of methods for content-based correction of the perspective distortion in camera-captured document images is presented in [5].

In this paper we present our ongoing research on document image mosaicing. We are inspired by the work in [17] for employing the inertial sensors for document mosaic image reconstruction. However, we are working on elaborating a lightweight algorithm that could be implemented on smartphones. The two novel contributions of our work are the following: (1) We use the gyroscope sensor to give visual user feedback to avoid perspective distortion during the video scan of the paper document whereas perspective distortion is not considered in [17]. (2) We compute the direction of swipe to optimize the keypoint feature descriptor-based image registration method by using only accelerometer sensor data whereas the authors of [17] optimize a keypoint feature descriptor-based image registration method by computing the displacement of the mobile phone from accelerometer and gyroscope sensor data.

The remainder of this paper is organized as follows. We present a detailed description of our method of video scanning of paper documents in Sect. 2. In Sect. 3 we discuss the experimental evaluation and the results. In Sect. 4 we present our conclusion and the future directions of research.

2   Mobile Phone Camera-Based Video Scanning

A block diagram of our method for mobile phone camera-based video scanning of paper documents is presented in Fig. 1. In this section we present a detailed description of our method for mobile phone camera-based video scanning of paper documents. We first describe the capturing of video frames and the recording of the accelerometer sensor data along with the gyroscope-based visual feedback for avoiding perspective distortion. This is followed by a description of our methodology for finding the direction of swipe (of video scanning), from the accelerometer data recorded with the captured frames in the video sequence. Finally we describe the image registration of the frames of the captured video and the reconstruction of the complete mosaic image of the paper document.

2.1   Video Scanning of Paper Documents

The video scanning of a paper document is achieved by a one-dimensional swipe of the mobile phone camera on the paper document. The swipe could either be from the top to the bottom of the document or from the bottom to the top of the document. During the video scanning we record the accelerometer sensor data along with capturing of the video frames.

Processing of Accelerometer Sensor Data: An accelerometer in a smartphone measures the acceleration $\alpha = (\alpha_x, \alpha_y, \alpha_z)$ of the phone in each direction of the X, Y and Z-axis. The accelerometers in smartphones are usually not of very high quality (because of cost constraints) and thus the obtained acceleration data is very noisy. Hence, the raw accelerometer readings are full of random noise and are unusable in their original form. The rise in temperature of the mobile phone (resulting from camera and screen heat) increases the random noise in the accelerometer data [16]. In order to make sure that the accelerometer reading is as close to the real value as possible, we compute the calibration offset of the accelerometer sensor by placing the phone on a flat surface and averaging x readings along each of the three axes separately and independently. We then subtract the calibration offset from the future readings of the accelerometer; hence obtaining calibrated readings. For removing random noise from the

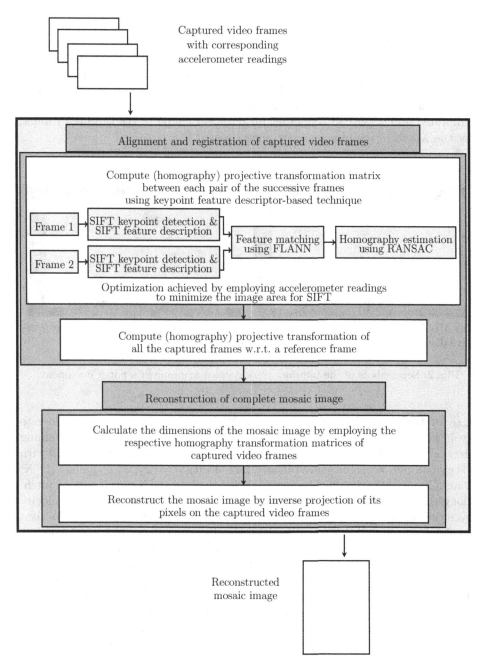

Fig. 1. Block diagram of method for mobile phone camera-based video scanning of paper documents.

accelerometer readings we smooth the accelerometer readings by using a Kalman filter-based running average. The smoothed accelerometer readings are recorded with the video frames. Figure 3 presents respectively the raw (noisy) accelerometer readings for the whole video scanning of a paper document, the calibrated and smoothed accelerometer readings for the whole video scanning of a paper document, the raw accelerometer data associated to the captured frames and the calibrated and smoothed accelerometer data associated to the captured frames, for the X-axis of the phone accelerometer sensor.

(a) Misaligned triangles result in perspective distortion. (b) Well-aligned triangles avoiding perspective distortion.

Fig. 2. Screenshots of the interface during video scanning of a paper document.

Use of Gyroscope Sensor Data for Visual User Feedback: In order to avoid perspective distortion, we employ the gyroscope sensor data for obtaining the orientation of the mobile phone camera. Modern smartphones are fitted with 3D gyroscopes, which measure the angular velocity $\omega = (\omega_x, \omega_y, \omega_z)$ along the X, Y and Z-axis of the phone. The angular velocity along the X-axis is termed pitch, along the Y-axis yaw and along the Z-axis roll. We obtain the angular velocity from the gyroscope as an angle of rotation around each of the three axes. We then employ it for providing a visual feedback to the user in order to keep the mobile phone camera parallel to the document plane and so to avoid the perspective distortion. We show three triangles on the screen for the visual feedback to adjust the orientation of the device. The three triangles are mapped to the gyroscope data along the three axes, respectively. A change in orientation of the device along an axis, rotates its respective triangle on the screen. By aligning any two of the three triangles, the user can keep the mobile phone parallel to the XY, YZ or ZX plane. To illustrate it pictorially, the Fig. 2 presents some screenshots of the interface during the video scanning of a paper document.

2.2   Finding the Direction of Swipe by Using Accelerometer Data

The accelerometer sensor data measures the acceleration of the phone along the X, Y and Z-axis. We use the calibrated and smoothed accelerometer data for

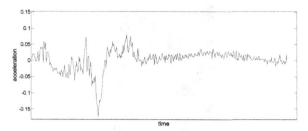

(a) Raw accelerometer readings of video scanning of a document.

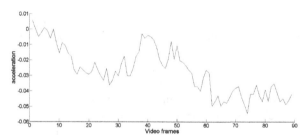

(b) Calibrated & smoothed accelerometer readings of video scanning of a document.

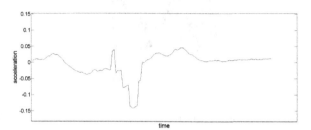

(c) Raw accelerometer readings associated with video frames.

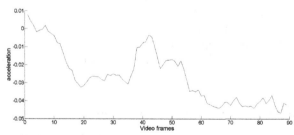

(d) Calibrated & smoothed accelerometer readings associated with video frames.

Fig. 3. Accelerometer data recorded for the X-axis of the mobile phone sensor during video scanning of a paper document.

Fig. 4. Windows phone accelerometer sensor coordinate system.

inferring the swipe direction of the mobile phone during the video scanning of the paper document. The accelerometer readings along each of the three axes is between $-1g$ and $+1g$ (where $g = 9.8\,\mathrm{m/s^2}$). The document page is placed on a planar surface (e.g. a table) and we use the phone in landscape mode (as shown in the screenshots of Fig. 2) for video scanning of documents. This setup makes the mobile phone's X-axis as the primary axis of swipe. The sensor coordinate system, indicating the X, Y and Z-axis of the mobile phone that we use for our research, is shown in Fig. 4.

To compute the swipe direction of the mobile phone (from the top to the bottom or from the bottom to the top of the document page) we employ a very simple methodology. We count the number of positive and negative readings in the recorded accelerometer data of the frames of a video scan. If there are more negative values than positive ones, this means that the phone is swiped in negative direction of the X-axis. And if there are more positive values than the negative ones, this means that the phone is swiped in positive direction of the X-axis. A swipe in negative X-axis direction corresponds to a top to bottom video scan of a paper document whereas a swipe in positive X-axis direction corresponds to a bottom to top video scan. This simple methodology is robust, efficient and very useful for detecting the direction of swipe during the video scanning of paper documents.

2.3   Image Registration of the Video Frames

The captured frames from video scanning of the paper document are registered by employing a keypoint feature descriptor-based technique. Successive frames in the video sequence are aligned and the projective transform or homography is computed between them. Afterwards, the computed homographies between successive frames are employed for calculating the homography of each frame to a reference frame in the video sequence.

Optimized Alignment of Successive Video Frames: For the alignment of successive frames in the video sequence, we use a keypoint feature descriptor-based alignment technique. As discussed by [15], the feature-based image alignment techniques have some very interesting advantages over pixel-based image alignment techniques. Namely for mobile phone camera-based video scanning of paper documents, the feature-based image alignment methods are more efficient and robust than the pixel-based methods; specially in case of uneven lighting and scene motions. There are many keypoint detectors and feature descriptors in the state of the art. They include the famous SIFT [9], SURF [2], FAST [12], ORB [13] and FREAK [1].

To align two successive frames, we first employ the SIFT keypoint detector to obtain a set of keypoints in two frames. Second, we extract the SIFT feature descriptors on each of the detected keypoints in the two frames. Third, we perform FLANN-based feature matching [10] between the two frames and employ RANSAC [3] for refining the initial correspondences obtained by FLANN.

In order to optimize keypoint detection, feature descriptor computation and feature matching, we use the direction of mobile phone swipe during video scanning. During the keypoint detection and feature descriptor computation phases we use the direction of swipe to avoid processing the complete image and to ignore the top and bottom parts of successive frames (or vice versa depending on the direction of swipe). As a result, it reduces the search space to be exploited during feature matching. The size of top and bottom parts of successive frames respectively to be ignored considering the direction of swipe is controlled by a parameter which is computed automatically for a video sequence and it takes into account the resolution of video frames (in pixels) and the speed of the swipe in the sequence (assumed to be constant during the video scanning). The speed of swipe is estimated from the total number of video frames in the video scan of a document page. The parameter for ignoring the top and bottom parts of successive frames is computed as:

$$G = \frac{h}{n} \tag{1}$$

where G denotes the number of pixel rows to be ignored on the top and bottom of successive frames respectively, h is the height of captured video frames in pixels and n is the total number of frames in video sequence.

Homography Computation for Two Successive Frames: We compute the planar homography or projective transform between two successive frames by minimizing the backpropagation error and further refine the computed homography by using the Levenberg-Marquardt method to minimize the backpropagation error [15]. The homography or projective transform between two successive frames is represented by a homography matrix.

For a captured video sequence of n frames given by:

$$V = \{f_1, f_2, f_3, \ldots, f_{n-1}, f_n\} \tag{2}$$

the set of homographies between successive frames is:

$$h = \{h_{(1,2)}, h_{(2,3)}, h_{(3,4)}, \ldots, h_{(n-1,n)}\} \tag{3}$$

where $h_{(i,j)}$ is the homography between frames f_i and f_j.

Homography Between Non-successives Frames: Employing the well established properties of matrices, the homographies computed for successive frames are employed in a cascade matrix multiplication, for computing the homographies between non-successive frames. For example for computing the homography $h_{(1,5)}$ between the frames f_1 and f_5, the homographies $h_{(1,2)}$, $h_{(2,3)}$, $h_{(3,4)}$ and $h_{(4,5)}$ are matrix multiplied. This permits us to define a homography between any pair of frames in the video sequence (whether successive or non-successive).

2.4   Reconstruction of the Complete Mosaic

The complete mosaic image of the video-scanned paper document is constructed by a projection of the pixels in captured frames onto a reference frame. Thus, we first select a reference frame in the captured video sequence. For simplicity, we suppose here that the first frame f_1 is selected as the reference frame. Then, we compute the size of the complete mosaic image by projecting the four corners of each frame in the video sequence using the corresponding homography matrix of the frame. To avoid holes or missing pixels in the complete mosaic image, the construction of the complete mosaic image is achieved by inverse projection of each of its pixels onto the sequence of frames. If a pixel of the mosaic is projected onto a subpixel in a frame, we use bilinear interpolation for computing the subpixel intensity from the intensities of the neighboring pixels. If a pixel of the mosaic is projected onto more than one frames of the video sequence we use the median value of the intensities of corresponding pixels of those frames.

3   Experimentation

In this section, we evaluate our method for mobile phone camera-based video scanning of paper documents on video frames captured at a resolution of 1280×720 pixels captured by a Nokia Lumia 920 smartphone. A custom application is developed for the capture phase of video scanning of paper documents. Some screenshots of this application are presented in Fig. 2. The video scanning application runs only in landscape mode to force the user to hold the smartphone with two hands and thus ensuring a stable orientation of the smartphone during capture.

During these preliminary experimentations the mosaic image reconstruction was performed on a laptop computer. The video scanning of paper documents was performed in an office environment with normal lighting conditions. The phone was kept parallel to the document plane by following the visual feedback on the orientation of the mobile phone camera, and the swipe was performed

Table 1. Number of frames captured during video scanning of the documents.

Page	Number of frames captured from video scanning at 1280 × 720 pixels
01	83
02	74
03	55
04	69
05	82
06	86
07	71
08	79
09	89
10	78
11	58
12	54
13	46
14	79
15	58
Mean	**71**

slowly and carefully. Table 1 provides the number of frames captured during the video scanning of the documents.

The experimentation dataset comprises fifteen A4-sized pages of scientific research papers; containing mostly printed textual content (in English). Some document pages contain also tables and mathematical equations. The document pages were printed on A4 pages and were digitized in three different modes, as given below:

1. image scanned by a classic scanner at 300 dpi grayscale
2. photo taken by the smartphone at a resolution of 1280 × 720 pixels
3. video scanned by the smartphone at a resolution of 1280 × 720 pixels

The image scanned by a classic scanner at 300 dpi serves as reference for evaluating the quality of reconstructed mosaics. We use the Levenshtein distance [6] as metric for comparing the OCR results of the images with ground truth. The Levenshtein distance between two string sequences is the edit distance between them i.e. the minimum number of single character edits (insert, delete, replace) required to change one string sequence into the other.

Table 2 presents a comparison of the Google Drive OCR results on documents for the three digization modes. The OCR results on the reconstructed mosaic from video scanning of the paper document at 1280 × 720 pixels, are better than the results on the single image captured at 1280 × 720 pixels. However they are not as good as those of the classical scanner scanned image. One important reason for this is that our method does not perform any camera calibration i.e. any wide angle lens correction on the captured video frames. The wide angle lens

Table 2. Experimental results

Page	# chars in page	Levenshtein distance between ground truth and Google Drive OCR results		
		Classical scanner image at 300 dpi	Single image at 1280×720 pixels	Mosaic from video scan at 1280×720 pixels
01	2569	36	1330	37
02	2311	44	808	57
03	1854	10	456	56
04	2353	10	441	125
05	2438	26	885	69
06	2495	22	1061	41
07	2171	20	678	542
08	2524	72	623	99
09	1422	296	785	354
10	2482	235	1417	487
11	2085	9	431	22
12	3286	382	3230	1442
13	4299	61	3786	111
14	3638	107	2112	1451
15	3924	300	3149	606
Mean	**2657**	**109**	**1413**	**367**

noise is thus inherited by the mosaic image and it eventually effects the OCR results. A second reason is that when the document pages were placed on the table for video scanning there was a small curvature at the corners whereas in case of a scanner this curvatures are flattened by closing the scanner lid. Our method does not perform any preprocessing of the document page.

Some examples of reconstructed mosaic images in grayscale from mobile phone camera-based video scanning of A4-sized paper documents are presented in Fig. 5.

Apart from the advantages of portability and liberty of scanning the documents of different sizes, the video scanning of paper documents is interesting as it allows to obtain the mosaic image at a higher resolution than the resolution of the camera used for capturing the frames. A mobile phone camera with a resolution of 1280×720 pixels can take a single image of an A4-sized page at 98 dpi as given by:

$$\text{dpi} = \sqrt{\frac{1280 \times 720}{8.27 \times 11.69}} = 98 \tag{4}$$

where, 8.27×11.69 is the size of an A4 page in inches.

Whereas when the same mobile phone camera is used for video scanning of A4-sized pages the reconstructed mosaic images have a much higher resolution (Table 3).

Table 4 shows a comparison of computation times for the reconstructed mosaics from video scanning at a resolution of 1280×720 pixels of A4-sized pages. Computation times are shown for mosaic construction with and without

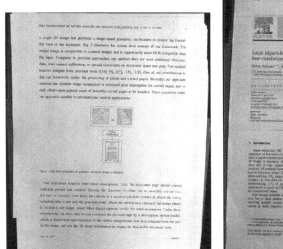

(a) Resolution:1389x1663 pixels, 153 dpi   (b) Resolution:1442x1761 pixels, 162 dpi

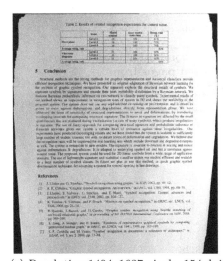

(c) Resolution:1404x1627 pixels, 154 dpi   (d) Resolution:1553x1659 pixels, 163 dpi

Fig. 5. Reconstructed mosaic images for A4-sized paper documents from video scanning at 1280×720 pixels.

Table 3. Resolution of the reconstructed mosaics from video scanning of the documents.

Page	Resolution	
	In pixels	In dpi
01	1442 × 1761	162
02	1409 × 1577	152
03	1389 × 1633	153
04	1507 × 1763	166
05	1427 × 1592	153
06	1405 × 1642	154
07	1456 × 1645	157
08	1479 × 1545	154
09	1381 × 1771	159
10	1346 × 1598	149
11	1369 × 1573	149
12	1553 × 1659	163
13	1432 × 1612	155
14	1404 × 1627	154
15	1467 × 1518	152

Table 4. Comparison of computation times

	Computation times (seconds) for mosaic construction of A4-sized pages from videoscanning at 1280 × 720 pixels	
Page	Without optimization	With optimization
01	419	401
02	433	333
03	244	223
04	355	336
05	381	369
06	419	401
07	346	330
08	377	361
09	391	375
10	365	338
11	264	251
12	266	259
13	258	257
14	440	421
15	289	278
Mean	**350**	**329**

optimized registration using accelerometer sensor data. Using the optimization a mean speed up of 21 s could be realized. We would like to highlight that the computation times are presented only to show the optimization of the classic feature keypoint-based image mosaicing. The times in Table 4 do not represent the best performance of our method for image mosaicing, since the code is not optimized yet and the mosaic construction was displayed on the screen step by step. The latter resulted into high times for mosaic reconstruction; both for optimized and non-optimized versions.

4   Conclusion

We have presented our ongoing research on the mobile phone camera-based video scanning of paper documents. Our method employs the gyroscope sensor of the phone for providing a visual feedback to the user for avoiding perspective distortion, and the accelerometer sensor of the phone for optimizing the keypoint feature descriptor-based image mosaicing technique. Our preliminary experimentation shows that the video scanning of documents not only allows to reconstruct the full page mosaic image of a document page from its mobile phone camera-based video scanning, but also reconstructs the full page mosaic image at a better resolution than the resolution of the camera lens used for video scanning. The work is in progress and we are working on the detailed experimental evaluation of the method along with an implementation on the smartphone platform.

Our ongoing research focus is on employing the gyroscope data for correcting perspective distortion of the frames in addition to the visual feedback. A second direction of ongoing research is to use super-resolution techniques for improving the quality of the mosaic image. In near future we will explore the use of the ambient light sensor for incorporating the lighting conditions of the video scan environment in mosaic reconstruction. In medium term we have planned to include a preprocessing step in our system for rectifying various geometric noises from mobile phone camera-captured document image frames.

Acknowledgment. The piXL project is supported by the "Fonds national pour la Société Numérique" of the French State by means of the "Programme d'Investissements d'Avenir", and referenced under PIA-FSN2-PIXL. For more details and resources, visit http://valconum.fr/index.php/les-projets/pixl.

References

1. Alahi, A., Ortiz, R., Vandergheynst, P.: FREAK: fast retina keypoint. In: International Conference on Computer Vision and Pattern Recognition, pp. 510–517 (2012)
2. Bay, H., Tuytelaars, T., Van Gool, L.: SURF: speeded up robust features. In: Leonardis, A., Bischof, H., Pinz, A. (eds.) ECCV 2006, Part I. LNCS, vol. 3951, pp. 404–417. Springer, Heidelberg (2006)

3. Fischler, M., Bolles, R.: Random sample consensus: a paradigm for model fitting with applications to image analysis and automated cartography. Commun. ACM **24**(6), 381–395 (1981)
4. Hannuksela, J., Sangi, P., Heikkila, J., Liu, X., Doermann, D.: Document image mosaicing with mobile phones. In: International Conference on Image Analysis and Processing, pp. 575–582 (2007)
5. Jagannathan, L., Jawahar, C.: Perspective correction methods for camera based document analysis. In: International Workshop on Camera-Based Document Analysis and Recognition, pp. 148–154 (2005)
6. Levenshtein, V.: Binary codes capable of correcting deletions, insertions and reversals. Sov. Phys. Dokl. **10**(8), 707–710 (1966)
7. Liang, J., DeMenthon, D., Doermann, D.: Mosaicing of camera-captured document images. Comput. Vis. Image Underst. **113**(4), 572–579 (2009)
8. Liang, J., Doermann, D., Li, H.: Camera-based analysis of text and documents: a survey. Int. J. Doc. Anal. Recogn. **7**(2–3), 84–104 (2005)
9. Lowe, D.: Distinctive image features from scale-invariant keypoints. Int. J. Comput. Vis. **60**(2), 91–110 (2004)
10. Muja, M., Lowe, D.: Fast approximate nearest neighbors with automatic algorithm configuration. In: International Conference on Computer Vision Theory and Applications, pp. 331–340 (2009)
11. Nakai, T., Kise, K., Iwamura, M.: Camera-based document image mosaicing using LLAH. In: Document Recognition and Retrieval XVI, pp. 1–10 (2009)
12. Rosten, E., Drummond, T.W.: Machine learning for high-speed corner detection. In: Leonardis, A., Bischof, H., Pinz, A. (eds.) ECCV 2006, Part I. LNCS, vol. 3951, pp. 430–443. Springer, Heidelberg (2006)
13. Rublee, E., Rabaud, V., Konolige, K., Bradski, G.: ORB: an efficient alternative to SIFT or SURF. In: International Conference on Computer Vision, pp. 2564–2571 (2011)
14. Sawhney, H.S., Hsu, S., Kumar, R.: Robust video mosaicing through topology inference and local to global alignment. In: Burkhardt, H., Neumann, B. (eds.) ECCV 1998. LNCS, vol. 1407, p. 103. Springer, Heidelberg (1998)
15. Szeliski, R.: Image alignment and stitching. In: Handbook of Mathematical Models in Computer Vision, pp. 273–292. Springer (2006)
16. Woodman, O.J.: An introduction to inertial navigation. Technical report 696, University of Cambridge, Computer Laboratory, Cambridge (2007)
17. Yang, Q., Wang, C., Gao, Y., Qu, H., Chang, E.: Inertial sensors aided image alignment and stitching for panorama on mobile phones. In: International Workshop on Mobile Location-Based Service, pp. 21–30 (2011)

Real-life Activity Recognition – Focus on Recognizing Reading Activities

Kai Kunze[✉]

Osaka Prefecture University, Sakai, Japan
kai.kunze@gmail.com
http://kaikunze.de

Abstract. As the field of physical activity recognition matures, we can build more and more robust pervasive systems and slowly move towards tracking knowledge acquisition tasks. We are especially interested one particular cognitive task, namely reading (the decoding of letters, words and sentences into information) Reading is a ubiquitous activity that many people even perform in transit, such as while on the bus or while walking. Tracking reading and other high level user actions gives us more insights about the knowledge life of the users enabling a whole range of novel applications. Yet, how can we extract high level information about human activities (e.g. reading) and complex real world situations from heterogeneous ensembles of simple, often unreliable sensors embedded in commodity devices?

The paper focuses on how to use body-worn devices for activity recognition and how to combine them with infrastructure sensing, in general. In the second part, we take lessons from the physical activity recognition field and see how we can leverage to track knowledge acquisition tasks (in particular recognizing reading activities). We discuss challenges and opportunities.

Keywords: Activity recognition · Reading · Cognitive tasks

1 Introduction

Activity recognition promises more pro-active computing assistants. If computing can recognize what we do during everyday life, it can actively support us in our tasks. Traditional context and activity systems focus on physical activities and rely strongly on specific sensor combinations at predefined positions, orientations etc. While this might be acceptable for some application domains (e.g. industry), it hindered so far the wide adoption of pervasive computing.

In recent years, physical activity recognition for very simple physical tasks has become relatively mainstream. industry begins to apply advances in activity recognition research, we see more and more commercial products that help people record their physical life, from simple step counting, over recording sports exercises, to monitoring sleep quality. Applying even newer results from research can help to recognize more complex tasks and make inference more robust.

M. Iwamura and F. Shafait (Eds.): CBDAR 2013, LNCS 8357, pp. 179–185, 2014.
DOI: 10.1007/978-3-319-05167-3_14, © Springer International Publishing Switzerland 2014

Fig. 1. Example for combining stationary and body worn sensors [1]

2   Robust Activity Recognition

Activity recognition has come a long way, from dedicated sensors in lab settings to being shipped in consumer products. For example, the new iPhone 5 s counts every step a user does (with the M7 motion co-processor) and tracks their mobility activity (walking versus driving etc.).

The initial systems only worked with a well-defined set of sensors on predetermined positions with known orientation. Yet, what happens if we want to use the recognition algorithms of these systems with commodity devices? How do placement variations of electronic appliances carried by the user influence the possibility of using sensors integrated in those appliances for human activity recognition? To overcome these problems, we designed a paradigm and classification for potential problems due to sensor displacement (informally also depicted in Fig. 2). We categorize possible variations into four classes: environmental placements, placement on different body parts (e.g. jacket pocket on the chest, vs. a hip holster vs. the trousers pocket), small displacement within a given coarse location (e.g. device shifting in a pocket), and different orientations.

For each of these variations, I give an overview of our efforts to deal with them [7]. We also describe initial research on how to dynamically combine environmental and body-worn sensors [1,13]. An example setup is shown in Fig. 1).

As we will see this principles more and more applied from industry, there are still a several open research questions related to activity recognition. First of all, how can a system combine changing sensor ensembles with heterogenous capabilities to provide reliable inference? We need a standardization for contextual information, defining lower level activities. For some cases, like modes

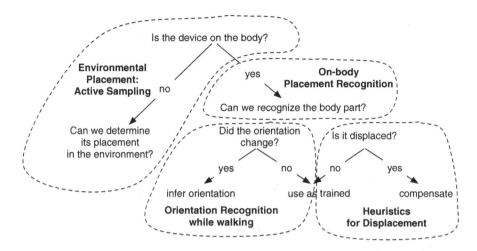

Fig. 2. Overview about the research questions and topics discussed in the first part of the talk.

of locomotion, this is easier solved (e.g. walking, running, standing) for others it's a very open question. Also on-body and environmental placement sensing are far from perfect and can be improved.

3 Towards Tracking Learning Tasks

There are several approaches possible if we talk about analyzing learning tasks. Sensing brain activity seems to be an obvious choice. Yet, doing this accurately demands invasive methods: electroencephalography (EEG), functional magnetic resonance imaging, and electrocorticography. EEG seems the most promising of those, as there are already some portable devices on the market. Yet, EEG is quite noisy and motion artifacts can overshadow the signal requiring additional processing and filtering.

Alternatively, we can track eye movements, which are strongly correlated with cognitive tasks. However, because eye movements include various types of information -about the users attentiveness, degree of fatigue, emotional state, etc.- it can be difficult to isolate the object of interest. There are two prevalent methods the uses optical eye tracking and using Electrooculography [5].

Sometimes other sensing modalities, like galvanic skin response or motion can give us insights into the mind. Yet, this is highly dependent on the tasks at hand. Also galvanic skin response seems to be quite user dependent.

Egocentric cameras are another interesting sensing modality. Although they are not able to detect cognitive tasks as such, they could be used to assess the stimuli a user faces or in the context of reading give a quantitative upper limit on the amount of reading a user can do (as all reading materials are seen by the camera). The same holds for capacitive sensors or other types of sensing focusing on physical activity.

As more and more learning and reading is done on digital devices, augmenting those devices can also give additional statistics. We can now track how often a user opens a given book ,when he underlines something, when he closes the book etc. These types of information are already gathered by Amazon, Google etc. and give interesting insights if combined with sensing technologies.

3.1 Reading Activities

We try to combine several pervasive sensing approaches (document image retrieval, motion-based activity recognition, eye tracking etc. Fig. 3) to tackle the problem of recognizing and classifying knowledge acquisition tasks with a special focus on reading [9,10,12]. Tracking reading enables us to gain more insights about expertise level and potential knowledge of users. We discuss which sensing modalities can be used for digital and offline reading recognition, as well as how to combine them dynamically.

Reading is interesting as it's a basic knowledge acquisition task, it's relatively easy to define what constitutes reading yet recognition is sufficiently complex, actually quite difficult if no eye tracking is used. We believe "Reading" in terms of cognitive tasks can be similar to "walking" or modes of locomotions in physical activity recognition. It's a first starting point that is challenging enough to get new insides about how to tackle more difficult tasks like comprehension.

As a first step to track reading, they implemented the Wordometer. Analogous to a pedometer counting the number of steps a user takes, the Wordometer estimates the words a user reads using the eye gaze recorded by a mobile eye

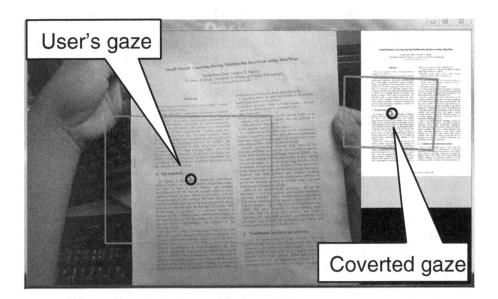

Fig. 3. Inferring the words read on a document by combining document image retrieval and mobile eye tracker data [10]

tracker and document image retrieval. We also investigate what you read using eye gaze. Tracking the document types gives more insights about expertise level and potential knowledge of users towards a reading log tracking and improve knowledge acquisition [11]. How often a user reads specific document types can provide insights into interests (e.g., comic versus belletristic) or language expertise and skills (e.g., computer vision textbooks versus English literature books).

An even more interesting question is whether one can estimate the text comprehension and the level of expertise of the reader using eye movements. In an initial study, they focus on assessing second language skills in students. Wearing the mobile eye tracker, the participants read several text comprehension sections from a standardized English test, answered questions, and highlighted difficult words afterwards. Looking at the frequency of fixations, one can determine all difficult words marked by the user.

3.2 More General Knowledge Acquisition

Although we focus so far on reading as a fundamental knowledge acquisition task, the methodology we applied can be easily extended to other learning activities and to other cognitive tasks (as also described in [4,6,8]). Take comprehension level, one can not only try to assess the reading comprehension but also the comprehension of an diagram, picture, video or an arbitrary other stimulus. Of course, the processing, sensors and analysis might differ significantly. The same holds for the document type analysis, we can try classifying types of movies or other art pieces. Yet, how well these systems will work has to be seen. Combining the cognitive science findings from the lab with robust activity recognition opens a new insights into the human mind, its functioning and dependencies.

4 Challenges and Opportunities

We could identify 3 major issues that need to be resolved for cognitive activity recognition to take off:

1. Ground Truth: Even during tracking physical activities, researchers have sometimes trouble defining the ground truth for some tasks. Some like modes of locomotion seem fairly easy to describe, others are more difficult e.g. consider "greeting somebody", this can be a fairly difficult context to describe and might be interpreted differently depending on cultural background. Now if we regard cognitive tasks, the problem increases, as it is in general hard to figure out what goes on in somebodies mind. For example, how do you define the ground truth for the reading comprehension? Even if you assess the level of understanding using questions or similar, somebody who did not understand the test at all might score well because of previous knowledge in the topic.
2. Sensing technologies: Right now there are very few mobile setups to track cognitive recognition. Usually they just use a single modality, e.g. mobile eye

trackers and eeg devices. So far there's also very little research in doing multi-modal cognitive task recognition. Defining and finding new sensor modalities and the combination of them will be hot topics in the future.

3. Privacy: Trusting somebody with our digital communications and physical activity logs is already challenging. Yet, trusting somebody with our reading log or our current comprehension level is even more challenging. How to safely store and process this data without violating privacy and ethical issues is very critical.

Despite these problems, the field offers a huge set of opportunities. We picked 4 interesting merits in the following.

1. "Quantified" Learning: We don't need to rely on few talented teachers who inspire their pupils. We can now better understand what type of learning might work. Questions that could be answered are: How much reading is necessary on average to understand this concept? Which are the most efficient ways to understand a particular topic? Which students/classes are good at a topic? Where are they lacking? It gets also interesting on an individual level, as we can now see more about our reading habits and get details about how well we are performing compared to colleagues, friends, fellow students.

2. Assessing requirements and dependencies in the large: Moving cognitive task monitoring away from the Lab, enables us to better assess secondary effects on learning. What are healthy sleeping cycles for learning? Are there types of food or other living circumstances that are beneficial to mental fitness? For which types of problems are discussions the right approach? Where is reading better?

3. Interactive way of storytelling: Initially outlined by Biedert [3] just for eye tracking, we can tailor story telling towards the user, by observing the mental state and displaying content accordingly: Making text passages easier or more dense to read depending on the current capacity and fitness of the reader. We can also introduce a video or graph when we recognize the user gets bored by the content and is close to drift away. Enabling cognitive tracking will revolutionize the way we tell stories.

4. User-Centric Document Analysis: this technology is not only for content consumers but also for content providers. Aggregating the cognitive experiences of users, we can tag documents with this information giving feedback to the authors. For example, a lot of users stopped reading after this paragraph. Most users were lost during this section and were really entertained by this part.

5   Conclusion

We outlined the emerging field of cognitive activity tracking by looking first at traditional physical activity recognition and its robust implementation. We summarized our efforts so far about a small part in this new research field, dealing with knowledge acquisition focusing on reading. We identified reading as a good start for exploring this field also over learning towards more general cognitive activities. Finally, we described where we see challenges and opportunities for future research. Let's combine physiology, cognitive science and wearable computing technology to figure out more about our minds and give us new insights about the functions of our brains.

Acknowledgments. This work was supported in part by the CREST project "Creation of Human-Harmonized Information Technology for Convivial Society" from the Japan Science and Technology Agency (JST).

References

1. Bahle, G., Lukowicz, P., Kunze, K., Kise, K.: I see you: how to improve wearable activity recognition by leveraging information from environmental cameras. In: 2013 IEEE International Conference on Pervasive Computing and Communications Workshops (PERCOM Workshops), pp. 409–412 (2013)
2. Beymer, D., Russell, D., Orton, P.: An eye tracking study of how font size and type influence online reading. In: Proceedings of the British HCI, pp. 15–18. British Computer Society (2008)
3. Biedert, R., Buscher, G., Dengel, A.: The eyebook-using eye tracking to enhance the reading experience. Informatik-Spektrum **33**(3), 272–281 (2010)
4. Bulling, A., Roggen, D., Tröster, G.: Wearable EOG goggles: seamless sensing and context-awareness in everyday environments. J. Ambient Int. Smart Environ. **1**(2), 157–171 (2009)
5. Bulling, A., Ward, J.A., Gellersen, H.: Multimodal recognition of reading activity in transit using body-worn sensors. ACM Trans. Appl. Percept. **9**(1), 2:1–2:21 (2012)
6. Bulling, A., Ward, J.A., Gellersen, H., Tröster, G.: Eye movement analysis for activity recognition using electrooculography. IEEE Trans. Pattern Anal. Mach. Intell. **33**(4), 741–753 (2011)
7. Kunze, K.: Compensating for on-body placement effects in activity recognition. Ph.D. thesis, University Passau (2011)
8. Kunze, K., Iwamura, M., Kise, K., Uchida, S., Omachi, S.: Activity recognition for the mind: toward a cognitive "quantified self". Computer, 98–101 (2013)
9. Kunze, K., Kawaichi, H., Yoshimura, K., Kise, K.: Towards inferring language expertise using eye tracking. In: Ext. Abs. CHI 2013, pp. 4015–4021 (2013)
10. Kunze, K., Kawaichi, H., Yoshimura, K., Kise, K.: The wordmeter - estimating the number of words read using document image retrieval and mobile eye tracking. In: Proceedings of the ICDAR 2013 (2013)
11. Kunze, K., Utsumi, Y., Shiga, Y., Kise, K., Bulling, A.: I know what you are reading: recognition of document types using mobile eye tracking. In: Proceedings of the 17th Annual International Symposium on International Symposium on Wearable Computers, pp. 113–116. ACM (2013)
12. Kunze, K., Yuki, S., Ishimaru, S., Kise, K.: Reading activity recognition using an off-the-shelf eeg. In: Proceedings of the ICDAR 2013 (2013)
13. Kurz, M., Hölzl, G., Ferscha, A., Calatroni, A., Roggen, D., Tröster, G., Sagha, H., Chavarriaga, R., del R. Millán, J., Bannach, D., Kunze, K., Lukowicz, P.: The opportunity framework and data processing ecosystem for opportunistic activity and context recognition. Int. J. Sens. Wirel. Commun. Control (Special Issue on Autonomic and Opportunistic, Communications), 102–125 (2011)

Author Index